T5-ASL-850

Internationalization of Research and Development

Springer
Berlin
Heidelberg
New York
Barcelona
Budapest
Hong Kong
London
Milan
Paris
Singapore
Tokyo

Klaus Brockhoff

Internationalization of Research and Development

With 20 Figures
and 7 Tables

 Springer

Prof. Dr. Klaus Brockhoff
University of Kiel
Institute for Research in Innovation Management
Olshausenstraße 40
24098 Kiel
Germany

338.926
B864i

ISBN 3-540-64819-4 Springer-Verlag Berlin Heidelberg New York

Library of Congress Cataloging-in-Publication Data
Die Deutsche Bibliothek – CIP-Einheitsaufnahme
Brockhoff, Klaus: Internationalization of Research and Development / Klaus Brock-
hoff. – Berlin; Heidelberg; New York; Barcelona; Budapest; Hong Kong; London; Mi-
lan; Singapore; Paris; Tokyo: Springer, 1998
   ISBN 3-540-64819-4

This work is subject to copyright. All rights are reserved, whether the whole or part
of the material is concerned, specifically the rights of translation, reprinting, reuse of
illustrations, recitation, broadcasting, reproduction on microfilm or in any other way,
and storage in data banks. Duplication of this publication or parts thereof is per-
mitted only under the provisions of the German Copyright Law of September 9,
1965, in its current version, and permission for use must always be obtained from
Springer-Verlag. Violations are liable for prosecution under the German Copyright
Law.

© Springer-Verlag Berlin · Heidelberg 1998
Printed in Germany

The use of general descriptive names, registered names, trademarks, etc. in this publication does not imply, even in the absence of a specific statement, that such names
are exempt from the relevant protective laws and regulations and therefore free for
general use.

Hardcover-Design: Erich Kirchner, Heidelberg

SPIN 10689741   42/2202-5 4 3 2 1 0 – Printed on acid-free paper

JK

Foreword

"After having acquired Boehringer-Mannheim, Roche Holding Corp., Basel, the Swiss pharmaceutical company which is one of the biggest R&D spenders in the industry with a budget of 2.2 Billion $, intends to establish its sixth research center in Bavaria. This center will be charged with cancer research that until now is located at Nutley, NJ. Jonathan Knowles, the new R&D director, intends to restructure R&D and to grant more autonomy to the researchers. This involves relocating about 500 people, not only in Nutley but also in Kamakura, Japan and Mannheim, Germany. This is where Boehringer has located its research. The other centers in Basel, Switzerland, Palo Alto, CA, and Welwyn, Great Britain, will keep their established fields of competence. Unlike in earlier years, Roche plans to have basic research in all of its centers." This message was printed in early 1998, about the time when the manuscript for this publication received its final touches. It explores many of the problems that we intend to deal with. Roche has internationalized its research and development. As a consequence of the acquisition of Boehringer-Mannheim, it needs to integrate the acquired research and development activities with its original activities. Environmental influences as well as the internal distribution of competencies lead to a redefinition of the missions assigned to laboratories. Probably the expensive relocations of personnel are necessary to finalize these plans. But, is all of this the best possible solution? Probably, nobody can answer this question. Too many uncertainties are involved, and too little is known about international research and development.

This study tries to collect knowledge on international research and development in firms as documented in the literature and a number of original studies that were performed by the Institute of Research in Innovation Management at the University of Kiel during recent years. We cannot claim final answers, and very often we can only present a collection of arguments that need to be weighed to make decisions. Thus, we demonstrate that decision problems in this area are

University Libraries
Carnegie Mellon University
Pittsburgh, PA 15213-3890

optimization problems as in many other areas. Choosing solutions that stress only one side is bound to lead astray. In some instances, earlier views have been corrected, and in some other cases, they have been refined. We hope that our findings are of interest to practitioners and that they spur interest of business researchers for their further development, refinement or reversal.

Most of our studies originate from a doctoral program in Management of Technology and Innovation. It was funded jointly by the "Deutsche Forschungsgemeinschaft" and the State of Schleswig-Holstein. Bernd Schmaul and Justus Bardenhewer were among the students who concentrated their work on international research and development. Other PhD-candidates, like Ute Pieper and Jan Vitt, looked into consequences of mergers and acquisitions for research and development, although not necessarily at an international scale. Christian Süverkrüp studied acquisition motives in cross-border acquisitions and considered those cases in particular where accessing new technology was of particular importance for the acquiring firm. Sven Vanini and Mathias Rüdiger are considering issues of knowledge management. Here, we tried to put their contributions into a unifying context.

Many of these dissertation projects were jointly supervised by my friend and colleague, Jürgen Hauschildt, and myself. He was patient enough to join in many discussions on the emerging text, and he offered advice. It is a pleasure to thank many colleagues and the PhD-candidates for the stimulating research atmosphere that they all created during the past few years. Subject to the rules of such programs, the funding comes to an end after nine years. By then, we shall have completed about 70 research projects, most of which are empirical. Thanks go to our sponsors and their reviewers who allowed us to dig deeper into a most fascinating field of research.

Also, our partners in industry who contributed their time by answering questionnaires, responding positively to interviews or joining in expert panels should be thanked. It is through them that we have been able to better understand a great number of problems. We hope to return some of their graciousness by offering our results in many publications and in this little book in particular.

Ms. Doerte Jensen was our expert for typing and editing a never-ending flow of manuscript revisions. We are proud to have a dedicated staff who supports our research work in such competent ways. Sven Vanini carefully read the text and suggested many improvements. Ms. Erika G. Niestroj-Frost revised the English in a careful way. To all of them go sincere thanks.

As always, the author resumes responsibility for all errors or unclear parts of text. Choosing a language that is not the author's mother-tongue is no small barrier to making oneself understood. Hopefully, not too many problems arise from this fact.

Institute of Research in Innovation Management
University of Kiel, Germany
August 1998   Klaus K. Brockhoff

Contents

1. Introduction and overview

In earlier years, industrial research and development (R&D) laboratories were exclusively located at a firm's headquarters. One company in the chemical industry even had two central laboratories serving different groups of strategic business units at its headquarters, before disbanding the concept of central laboratories almost totally and adopting a concept of internationally dispersed R&D investment centers. Today, even within one industry and among the biggest global players, we can observe firms which have spread their R&D laboratories all over the globe as well as those which have centralized these activities. For instance, we can observe both of these phenomena in the chemical industry.

A substantial number of reasons could be named for the original centralization of R&D. The perceived need for a critical mass of researchers combined with the hope for economies of scale, high costs of communication and coordination between geographically dispersed units, and an easier way of controlling the R&D portfolio as well as proprietary information work in favor of centralization. Perhaps even more important, firms found it beneficial to exploit firm-specific technological advantages that were offered by public research institutions in their home markets. Product life cycles extended long enough to allow the transfer of new knowledge from such institutions into firms via hiring young researchers and only occasional participation in R&D cooperations. Public centers of R&D competencies were often supported continuously by firms to build long-term relationships which again helped to ease the transfer of personnel. Today, none of these reasons remains undebated, and few remain valid.

This might explain why even the casual reading of press releases or annual reports of major technologically sensitive companies reveals that these firms have established R&D laboratories in many countries. Whether this occurs as a consequence of internationalizing marketing and production activities - which is a view advanced by some researchers on multinational firms (Hewitt, 1980) - or whether it reflects a strategic move of its own - which appears to be a more recent view - can hardly be established.

A further perspective is of political nature and concerns the environment in which firms want to do business. Each national environment might be characterized by specific legal or economic constraints on activities of foreign firms. For example, demanding "local content" is no longer restricted to production. It could also be aimed at R&D. In view of technological advances brought about by a firm's laboratories, it is asked how the worldwide distribution of laboratories affects the development of nations (Ajami/Arch, 1990). The internationalization of private R&D has become of such importance that public policy measures to deal with it are discussed in some of the literature (Granstrand/Håkanson/Sjolander, 1993). We do not take this view. Rather, we approach the problems of R&D internationalization from the economic perspective of a single firm.

Under this particular firm-specific perspective, it is necessary to discuss conditions for the existence of more than one laboratory within a firm. Only if the feasibility of many laboratories can be shown does it make sense to discuss the location of some of them in countries other than the home of the headquarters. These theoretical .arguments are developed in Chapter 2.

In the following chapters we deal with issues of internationalization, primarily in the sense of a firm having laboratories in countries that are different from the country where its home base is located.

At first, one might think that internationalization of R&D is a major topic in the literature on multinational firms. It is found that this is not so (Gerybadze, 1997, p. 29). The few studies which treat this issue empirically, indicate that firms can be more or less active in internationalizing R&D as compared with other functional activities (Colberg, 1989, pp. 206 et seq.). There is no strong correlation between the foreign activity level of various activities. This is a first result showing that R&D activities have not simply followed in a proportionate manner to production or marketing activities. Studies on foreign R&D indicate that this function of firms is concentrated more geographically than production or trade activities (Florida/ Kenney, 1994; Dunning/Narula, 1995; George, 1995). Interaction with conditions in the host country and specific characteristics of producing new knowledge by R&D might explain this observation. The specific character of such conditions is explored in Chapter 3.

Disregarding the "hen or egg-question" of whether R&D or other business functions lead internationalization, we can observe that internationalization of R&D is a growing activity. This can be demonstrated by looking either at R&D output or input data:

(1) Patents serve as one set of output data. Patent data suggests that U.S. firms have not expanded their internationalization recently (Patel, 1996) while Swedish multinational firms have not only spread their technological activities to more and more classes of technology in the post-war period, but their foreign laboratories have also contributed an ever increasing share of their patent activities (Zander, 1997). If this is true for firms in other countries as well, it could have two interpretations. First, a market driven increase of complexity requires firms to enlist the help of more and more researchers and technicians that could not be found at the home base. Secondly, the supply of more and more R&D results in places separate from the "traditional" centers of excellence in research necessitates a broadening of local contacts. It is argued that national systems of R&D have adopted a polycentric structure of knowledge generation around the globe that might be exploited by deliberate global sourcing activities (Gerybadze, 1997, pp. 18, 25). This opportunity was used primarily by multinational firms, and it led in turn to more complex products. Whether the direct between the greater spread of knowledge producing institutions over the globe is more important as an influencing factor on R&D internationalization than the indirect link via product complexity is yet unknown. Both explanations support an increase in foreign R&D activities. Interestingly, the explanations seem to negate recent drives towards a concentration of product portfolios and product development on core technologies.

(2) Let us now consider R&D input data. Inference on the early developments of foreign R&D activities by private firms can be drawn from a number of questionnaire studies that were performed in the U.S. and some European countries. However, these studies are not representative for the whole industry in the respective countries. In addition, expenditure data and head counts might signal developments that are not quite correlated

(Brockhoff/von Boehmer, 1993, p. 401). Nevertheless, it is concluded again that the internationalization of R&D is a growing phenomenon (von Boehmer/Brockhoff/Pearson, 1992; Granstrand/Håkanson/Sjolander, 1993; Beise/Belitz, 1996). It is argued that a need to organize global systems of technology acquisition exists, because impulses for innovation as well as competencies for R&D are distributed globally and that it might be more economical to use resources in foreign countries.

This view is supported by reading national statistics which provide information on international R&D expenditures by national companies, sometimes collecting only those expenditures that exceed a certain amount. Examples from Switzerland or Germany can be quoted (Buri/Suarez de Miguel/Walder, 1986; Deutsche Bundesbank, 1996). In 1995, German companies spent about 10 Billion DM or 17% of their R&D expenditures in the same year on R&D abroad. The major share of these foreign R&D expenditures resulted from the chemical industry, and its major destination was in U.S. laboratories. An almost equal stream of payments (9.5 Billion DM) was channelled from foreign companies into Germany with almost equal shares originating in the U.S. or other European countries (BMBWFT, 1998, pp. 45 et seq.). Swiss companies spent about 20% more on foreign R&D than on domestic R&D (Vorort, 1998).

In Japan, information on foreign R&D expenditures and foreign R&D establishments by Japanese firms is collected by a number of agencies. MITI has published a summarizing report (Ministery of International Trade and Industry of Japan, 1995). In Japan, internationalization of R&D is a relatively recent phenomenon and has not reached the level found in other industrialized countries. The recency of internationalization of R&D can be concluded from a questionnaire study which reveals that the median year for starting international R&D by Japanese firms was 1986, while 43% of 206 laboratories established outside of Japan were founded in 1990 or later (Kurokawa/Iwata, 1997).

In the U.S., the National Science Board has published a time series on foreign R&D in the U.S. as well as R&D performed abroad by U.S. firms. This information is classified by industry, nationalities, expenditures and establishments (National Science Board, 1996). According to this source, 10.2% of U.S. companies' R&D expenditures was performed abroad, while foreign sources financed 9.8% of

U.S. industrial R&D expenditures. In absolute numbers, the total is about 40 billion $.

For other countries, less representative and comparative information is found. While the existing statistics demonstrate the importance of international R&D, they provide only partial information on this topic. This has to do with two aspects of our topic. R&D has neither well-defined boundaries, nor is internationalization a term that is well understood. Let us consider these two aspects very briefly.

At first, in spite of attempts to standardize the term R&D for statistical purposes (OECD, 1993) there remain considerable "grey areas". This has also led to varying perspectives of empirical research. For instance, Behrman/Fischer (1990, p. 10) insist on having excluded technical service and quality control activities from their study. The spectrum of activities performed by a laboratory is not always clearly recognizable, and even if so, not all researchers deal with it in the same way.

Secondly, the term internationalization might cover a number of activities that are of a substantially different nature from the point of view of the management. Let us illustrate this point.

Internationalization of R&D could mean:

(a) Employment of foreign R&D personnel in a strictly national, possibly headquarters-based organization.
(b) Importing some new technological knowledge from a foreign country, for instance buying patents or licences.
(c) Engagement in cooperative R&D with partners located in a foreign country and involve some transfer of resources into a foreign country. This might occur as a project of limited duration and could take the form of a collaborative activity tied to a joint venture including production facilities (Behrman/Fischer, 1990, pp. 79 et seq.). Unlike the project, this arrangement is usually thought to be permanent. Certainly, these alternatives require substantially different arrangements with respect to sharing and accessing resources, choosing management and supervision structures, etc.
(d) Establishing a unit which performs some type of R&D work in a foreign country, where all projects under its supervision are exclusively performed within this unit (intralocal). It may be necessary to differentiate between units that are exclusively owned by one company or that are jointly owned by the company in ques-

tion and other owners,which could be companies, government institutions, universities etc.

(e) Performing interlocal R&D projects that make use of resources available in laboratories located in different countries, but which are fully owned by one company.

In this presentation, after briefly touching on some aspects of case (c), above, we are mainly concerned with situations that are best described by (d) and (e) in the foregoing classification. Particular management problems arise from these situations that are not present in more transient cases. The establishment of foreign R&D units might involve major investments, and it is mostly thought to be of a permanent nature. The decisions that lead to it will, therefore, involve very careful and deliberate reasoning. In Chapter 4 we plan to discuss alternative ways of establishing R&D laboratories.

After accepting the advance of internationalization of private R&D as a fact, one likes to learn why it occurs. At first, placing some R&D abroad may have appeared as an act of despair of those multinational high-technology firms that happened to be headquartered in relatively small countries, such as the Netherlands, Sweden or Switzerland. In such countries demand for qualified R&D personnel could not always be satisfied locally. Firms in these countries could only grow by exporting relatively high shares of their outputs. The need to adapt products to the demands of customers in other countries soon became apparent. Thus, one could argue that firms in such environments were forced to internationalize if they wanted to protect their competitive position and to continue with their growth. During recent years companies from much larger countries are following the same route, and these seem to do so voluntarily. "Although the reasons for this ... are complex, generally it appears that multilateral, industrial R&D efforts are a response to the same competitive factors affecting all industries: rising R&D costs and risks in product development, shortened product life cycles, increasing multidisciplinary complexity of technologies, and intense foreign competition in domestic and global markets" (National Science Board, 1996, p. 4-43). A further reason is that communication costs have declined which makes it easier to decentralize. A more strategic perspective is taken by Behrman/Fischer (1990, pp. 18 et seq.). They posit that the intensity and the mission of foreign R&D activities is primarily a consequence of a firm's market orientation. To describe this orientation they use three categories: home-market companies, like extractive industries or offshore component manufacturers; host-market companies, with a primary orientation towards particular national markets as in food or chemicals; and world-market companies, with a geocentric management style and a centralized structure, trying to establish a world product. Once one of these market orientations is chosen, a particular R&D strategy is expected to follow suit with some moderation due to aspects such as market size or availability of skills. These aspects alter one strategic factor into a higher complexity of reasons.

In Chapter 3 we review the empirical literature about the pros and cons of the internationalization of R&D, and we discuss why internationalized R&D is different from many laboratories in a home country.

The trend towards higher shares of foreign R&D aggregates a few withdrawals and many new engagements at the same time. The withdrawals, in particular, indicate that the expectations that were tied to the decision of acquiring or setting up foreign R&D laboratories have not always been met. In some cases, the environmental conditions that supported the decision in the first place may have changed. Whether better foresight could have helped firms to avoid what in hindsight appears to have been a wrong decision should not be discussed here. In some other cases, it becomes obvious that management cannot solve the problems that are created by relocating laboratories to foreign countries. These are the cases that need to be studied. Results from such studies may contribute to improved management of R&D. These observations raise a large number of questions which are difficult to answer. As mentioned before, in Chapter 4 we considered the "birth" of international R&D. In Chapter 5 we deal with the assignment of tasks to R&D laboratories and their control.

To avoid unwanted duplication of effort in internationally dispersed R&D units as well as to optimize the effectiveness of their efforts, companies search for structures and processes that promise to be successful. The overriding problem here is that of coordination. During the past few years coordination by hierarchy, by hybrid mechanisms or by market forces have been topics that received a lot of attention. In Chapter 6, we deal with some of the considerations resulting from this research which have particular relevance for international R&D such as coordination across organizational interfaces that arise between internationally dispersed R&D units and their company environment. We also discuss to what degree modern communication technologies might help researchers and managers to overcome some of the apparent communication problems.

Key to the control of international R&D is the problem of measuring its success. Very few studies have been devoted to this topic. As will become clear, multiple perspectives, multiple success criteria, time lags and interaction of R&D with other company or external activities for bringing about results contribute to the difficulties of measuring success. No standard system has yet evolved from the different approaches to the problem. A short review of the state of the art in this area is given in Chapter 7.

The research on which we draw in this presentation is of different character. Some is theoretical, most of it is empirical. Much can be learned from company reports. They show problems and solutions, although these have mostly been adopted on grounds of plausibility. Such problems and ad hoc solutions are reported by a number of practitioners. Chester (1994) reports on Hughes Aircraft. Perrino and Tipping (1990) of ICI discuss an approach to formulating an R&D deployment strategy for a global technology network, which arises out of a study commissioned by a leading management consulting firm. Zaininger (1990) discusses some of the issues addressed by Siemens AG regarding its headquarters in Germany and a newly established research center in Princeton, NJ. Seiffert (1990) adds his view on the world-wide division of R&D activities of Volkswagen AG, some of which are in cooperation with other firms. Krogh (1990, p. xxxviii), Senior Vice-President for R&D, has stated that 3M has "a global committement to locating laboratory sites wherever the product or technology driving forces exist". The issue of coordinating worldwide R&D activities has been addressed by Granstrand and Fernlund (1978), who describe the structure and processes SKF had established to facilitate communication among its R&D units and the cost incurred from this. Farris and Ellis' (1990) survey of U.S. companies shows that globalization ranks second in importance as a change force and that respondents to their questionnaire believe that in the past they have handled this area of management the least effectively. Howells (1990, p. 133) argues that a number of recent trends within R&D are affecting the existing balance of organization and locational factors within research operations. These are "...the increasing scale, scope and specialization of research activity; ... the internationalization of R&D; the increase of interorganizational linkages in research; and ... associated with this the growth of computer-communication networks and on-line-information systems on R&D". Erickson (1990) states that "as global networks form, companies will respond to forces pushing for greater efficiency of scale - for global perspective," and argues that R&D is one of the areas in which attempts have been made to achieve higher efficiency and more globalization through mergers and joint ventures. Taggart (1991), studying the pharmaceutical industry, observes the importance of accessing technology around the world.

To date, little information is available that summarizes the knowledge and the controversies on the various aspects of R&D internationalization that were mentioned above. In the following chapters we attempt to present such information. However, at first it is necessary to explore why companies should operate multiple R&D units and what makes international R&D specific as compared with national laboratory sites.

2. Multiple locations for R&D

Before trying to tackle problems of internationalization it might be good to answer the question why a company may want to entertain more than one R&D laboratory at different locations. The problem of internationalization in the sense of assuming a headquarters laboratory and at least one more laboratory in a different country, would then appear as a generalization of the first problem.

In answering this question, we will concentrate on economic criteria that might explain the observation. Certainly, non-economic reasons play a role as well in answering this question. Powerful regional managers, emotional relations with regions that are claimed to provoke innovative ideas, or other considerations may lead to establishing laboratories that otherwise might not have sprung up. However, unless these reasons appear to be tied to economic criteria, we do not intend to consider them in the first place.

Economic criteria are in some way related to the costs and benefits of the operations. In the extreme, infinite costs arise if certain input factors are not available in locations where laboratories are planned or exist. Usually, the internal costs of generating the desired outcome from R&D need to be considered. As R&D results might require inputs from other R&D performing institutions outside the respective firm, transfer costs that arise from the knowledge exchange between these institutions need to be considered. Similarly, transfer costs can result from passing on R&D results to internal or external "customers" further down the value chain.

With respect to the potential benefits of R&D, it should be kept in mind that R&D results are not meant to be repetitive. This is completely different from mass production. It means that each and every project is started with the intention to generate a result that was not known before, and projects may, therefore, require combinations of input not used in earlier projects. Potential benefits from R&D can be of different types. Firstly, there is the immediate benefit that arises if R&D results can be transferred into successful innovations or if they can be traded in for some other valuable resource. Secondly, R&D activities might help to identify valuable external knowledge and eventually to absorb it for the purposes of the company in question. This is R&D's absorptive capacity (Cohen/Levinthal, 1990). Here, we are not as concerned with the benefits of R&D as with its cost implications.

We will now try to identify reasons for multiple laboratory locations within one company. One way of doing this is by rejecting arguments that seem to favor the concentration of R&D in one location (see (1) and (2), below); another way is to look for arguments that favor the dispersal of these activities (see (3), below). The first approach makes it necessary to distinguish between unrelated and related R&D projects. Viewing the arguments collectively one arrives at the idea of optimum laboratory size (see (4) and (5), below).

(1) If projects in a laboratory are technically totally unrelated, then two arguments could be raised for putting them in one location. First, it might be argued that synergisms may result from a joint use of administrative and infrastructural inputs. The latter argument implicitly assumes that neither infrastructure can be partitioned into smaller units nor can excess capacity cannot be used by some other business function or sold to external parties. Alternatively, to support this argument one needs to prove that there are economies of scale in using the infrastructure. Here again, alternative uses could be considered. With respect to administrative inputs this argument does not become much stronger. During recent years, in particular, it was learned that such routine activities could be outsourced which means that a market price for any level of such required services exists. In this case, no synergism can result from having these services performed in-house.

Second, choosing only one laboratory location while its internal customers are internationally dispersed causes increasing communication costs which are not solely the result of a higher cost per unit of communication charged for the longer distance of travel. Primarily, increased communication costs reflect the observation that efforts to maintain a certain level of communication grow as the distance is increased. The complementary observation is: frequency of communication decreases almost exponentially with increasing distance (Allen, 1977). Modern communication devices might shift the level of frequency but do not counteract this trend. Thus, new information devices continue to call for close customer-laboratory distances. Assuming a wide dispersal of customers, these distances do not necessarily support one laboratory location.

(2) If R&D projects are related to each other it could be argued that experts may produce synergism from an exchange of ideas which could grow as more experts are assembled in one place. Evidence exists that lateral spillovers between R&D teams contribute to the results of R&D projects (Bergen, 1990). Furthermore, the response of one large laboratory to a fluctuating demand for its services could be more flexible than is expected from a number of smaller units that, in sum, represent a comparable capacity.

However, it should also be noted that such benefits come at a cost. As the number of experts grows, the possible number of interrelationships grows even more. Reducing our reasoning to binary relations we find that among n experts there can be n(n-1)/2 such relations. These need to be coordinated as the projects are interrelated. Therefore, the costs of coordination need to be weighed against the potential benefits.

(3) Highly qualified human labor is a major input factor for R&D. The desired level of qualification may not be available at the quantity needed at the traditional laboratory site. Transferring these factors from their present location to the laboratory may involve costs that are much higher than offering a more convenient location to the more or less immobile experts. This is particularly so in countries with a very heterogeneous level of regional development, such that attractive and unattractive regions result. For instance, production facilities might be located in an unattractive region because a necessary raw material can be found in this area or because environmental standards would not permit the same production in another area. Exactly the same reasons may keep away R&D specialists, who are offered more attractive alternatives. Thus, an upper constraint on size might exist.

Interestingly, a lower constraint on size or a minimum laboratory size (Mansfield/Teece/Romeo, 1979, p. 190 et seq.) is not very likely. The theoretical arguments against this view use the possibility to partition a R&D task, such that there is no "naturally" given task size, and they draw on the observation that input factors can be partitioned as well (at least to the degree that it is of interest here). It was F. M. Scherer who observed once that the most important input in knowledge production, creative thinking, can be gained from the human factor of production which comes in "man-sized lumps" (Scherer, 1971, p. 356). Empirical data from Germany shows that if a minimum laboratory size can be established at all, its employment capacity is much smaller than advocates of the minimum size argument had asserted (Brockhoff, 1994, p. 81 et seq.).

Weighing the different arguments one might conclude that neither one nor numerous laboratory sites are optimal. Rather, there exists an optimal laboratory size that in turn determines the number of laboratories if some volume of tasks is given.

(4) Beckmann/Fischer (1994) assume a n-shaped relationship between a laboratory's size and its economic performance. They discuss rather broad ranges of employment as the independent measure for less than critical and more than critical size. However, the dependent performance measure is not operationalized at all. If this relationship could be demonstrated empirically, then it would offer a very good explanation for the observation of multiple laboratory sites within one company.

Taking perceptions of laboratory performance as the dependent variable, Kuemmerle (1997) demonstrates a nonlinear relationship between laboratory employment, as a size measure, and performance. From this we can deduce an optimum laboratory size. This optimum laboratory size appears to be remarkably small, approximately 235 people for all laboratories. The optimum size of the laboratories in the pharmaceutical industry is 167 people, while it amounts to 254 people in the electrical/electronics industry. Also, development laboratories appear to have a larger optimum size (260 people) than research laboratories (182 people). It is also of interest that the optimum size of Japanese laboratories (211 people) has a relatively low value. This could be a result of the relatively higher costs of more interaction in group decision-making in Japan as compared with a more individualistic type of R&D decision-making in "Western" countries.

(5) In considering an optimum laboratory size, it is valuable to differentiate between the costs of inventing and the costs of communicating inventions. As argued above, communication costs depend on distances. With that in mind we can construct a very much simplified model.

We assume a u-shaped cost curve $c_d(x)$ by which a given stream of inventions can be generated from laboratories of size x. The optimum laboratory size x^* is the one for which $c_d(x^*) = \min_x c_d(x)$. In addition to the costs of invention we need to consider costs of communication. Let us assume that this depends on the level of the stream of inventions. As this level is assumed to be constant, so is the communication cost. Communication might arise with local partners only or with a mix of local and foreign partners. If all communication is with local partners of a laboratory, then we have communication costs c_l. If a laboratory is involved in communication with a foreign site as well, communication costs rise to $c_f > c_l$.

Now, assuming that one laboratory exists at optimum size and that it needs to communicate to local and foreign sites, the total costs of invention and communication would be $c_d(x^*) + c_f$. If, instead, two laboratories of sub-optimal size, e.g. $x^*/2$ each, were established at the respective location of their communication partners, the total costs would be $2 \cdot c_d(x^*/2) + c_l \cdot$. This could be advantageous under a cost-minimization objective if $c_f - c_l > 2 \cdot c_d(x^*/2) - c_d(x^*)$.

For the sake of illustration we assume that:

$$c_d = \frac{x}{625} + \frac{100}{x}.$$

This leads to an optimum laboratory size $x^* = 250$ at $c_d(x^*) = 0.8$. For all $c_f > c_l + 1.2$ it would pay to run two laboratories rather than one[1]. If an additional cost of coordination would arise, then this would have to be taken into account as well. The communication cost difference is very considerable relative to the costs of invention. It is not very plausible to assume such a large difference. However, it is more reasonable to question one of the assumptions underlying this model. Having two laboratories rather than one might be more easily rationalized by the inter-

---

[1] We find $c_d(x^*/2) = 1.0$, such that $2 \cdot 1.0 + c_l < 0.8 + c_f$ implies $c_f > c_e + 1.2$.

action between the size of the stream of inventions to the closer relationship that the two laboratories can have with their "customers". This effect needs to be balanced with eventual synergism arising from joint activities in one laboratory.

To generalize these few observations they indicate that companies are well advised to consider establishing more than one laboratory. The actual dispersal of R&D activities might arise as the original unit surpasses its optimum size.

Figure 1: Optimum laboratory size

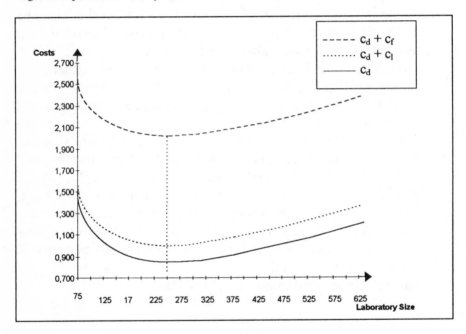

3. Particular aspects of internationalization

3.1 Universal success factors of innovation

One reason for the internationalization of the inventive phases of innovation processes could be that country-specific success factors exist. If relatively more efficient inventors would live in one nation and relatively less efficient ones in another, we should expect invention activities to be concentrated in the relatively more efficient places. Better education is thought to provide such differences in efficiency. Thus, education appears to be one of the success factors of innovation. However, we cannot assume that this factor is of importance in some nations and of no importance in some others. Rather, the level of a universally important factor is different. This view is supported in the extensive study on "Culture and Technical Innovation", where it is said that: "there is no evidence of any culturally different patterns of generating basic innovations" (Albach, 1993, p. 61) and "...we maintain that the determinants of success of technical innovation are the same all over the world. They are . . . competence, leeway, integration, commitment, and knowledge" (Albach, 1993, p. 78). These variables are then studied at the level of: the individual, the group, the firm and its environment as characterized by users, the organization of an economy, trade unions, education and the government.

It is apparent that the aforementioned variables might take quite different values in different countries. For example, it is found that four factors by which countries can be distinguished in regard to the values that managers or employees hold apply equally to R&D professionals. Differences in the level of each factor across countries can be substantial. One of the four factors is labelled "individualism" (with its opposite pole "collectivism"). With respect to this factor it is observed that: "... challenging work in a country high in Individualism may carry the meaning of individual achievement, responsibility, and control over outcomes, whereas in countries high in Collectivism it may mean contributing to the well-being of the in-group, showing loyalty, or achieving high status" (Hoppe, 1993, p. 319). One of the conclusions drawn from observations which show substantial differences in variable levels across countries is that country-specific management and organization models might have limited applicability for managing R&D in other countries. In particular, U.S. managers "with their highly individualistic and masculine 'baggage', and their emphasis on risk-taking, creativity, and levelling of power differentials, . . . often run counter to the values other countries hold" (Hoppe, 1993, p. 324). The collision, fragmentation, or separation of subcultures that might result from attempts to force a culture perceived as unattractive on R&D professionals.

If the inclination to accept another subculture is high and if this other subculture is perceived as being attractive, it could lead to an assimilation of this subculture (Cartwright/Cooper, 1993).

Another consequence of the observation of country-specific levels of variables is the initiation of a search of firms for those countries that offer a combination of value levels that are most favorable to the successful performance of the intended inventive processes. It is even thought that there exist some factors "which may be influenced by appropriate measures of company policy if they are recognized by management" (Albach, 1993, p. 120). This broadens the possibilities because the combination of two types of factor values, those accepted as being environmentally given and those that could be influenced to best support the objectives of a firm, need to be identified and used. As some of the so-called cultural factors can be influenced by economic forces, it is hardly surprising that "the share of success of innovations that can be explained by economic factors is larger than generally assumed" (Albach, 1993, p. 120). However, this is a result of qualitative reasoning and not the outcome of an econometric model that would have considered all rivaling factors simultaneously. Broadly based empirical research has shown that where competitive conditions are perceived in a similar way, managers tend to choose comparable strategies which are described by a set of variables and their values (Balachandra/Brockhoff, 1995; Bardenhewer 1998; Weisenfeld-Schenk, 1995).

The success factors mentioned above are those that help to bring about inventions. To what degree these are accepted by customers is yet a different question, although the factors governing acceptance might be influenced by activities of inventors.

It is uncertain whether national differences exist in tolerating innovations, in the predisposition to buy or to actually adopt innovations. Psychologists have not developed convincing evidence of such differences (Schulz-Hardt/Lüthgens, 1996). In a study on students' reactions to product preannouncements national differences can be observed (Schirm, 1995, pp. 152 et seq.): German subjects attribute significantly less credibility to preannouncements of revolutionary new products than students from other countries, in particular those from France. Students from Great Britain, Sweden and Hongkong attribute relatively more credibility to preannouncements by market leaders, which makes it more difficult for start-ups or niche producers to be credible. A broadly-based study on consumer innovativeness across eleven European countries finds differences measured by the "exploratory acquisition of products" scale (Baumgartner/Steenkamp, 1996). The authors find that: "innovativeness was positively affected by the importance a consumer attaches to . . . openness to change, and negatively . . . by the importance attached to conservatism. Innovativeness decreased with higher ethnocentrism, a more favorable attitude toward the past, and with age . . . Consumers in more individual countries tended to be higher in innovativeness . . . Further, innovativeness was found to be lower in cultures emphasizing uncertainty avoidance" (Steenkamp/Ter Hofstede/Wedel, 1997, p. 20). Whether these results could be extended to other regions and to industrial goods (for which a scale measuring innovativeness is yet to be developed) is not clear. However, there are indications of national differences in innovativeness that might explain the influences hypothesized above.

In the present context we are concerned only with factors leading to inventions via R&D activities. However, these aspects might not be totally unrelated. The relationships are sketched in Figure 2. Within one nation we might observe levels of success factors for inventions F_1. Should these lead to inventions, it is of interest to introduce these into the (national) market. Other factors and their levels F_2 will determine potential success of the resulting innovation. Successful inventions might influence the level of F_2 within one group of people. The willingness to accept innovations might have an influence on F_1 as well. For example, opposition to nuclear power plants, genetically manipulated food, or magnetically-elevated trains constitute unfavorable conditions for R&D. This opposition can cause a loss of competencies as well as commitment and, in turn, reduce inventive activities. Thus, our analysis of the factors of invention would be only a partial one.

Figure 2: A simplified model of invention and innovation

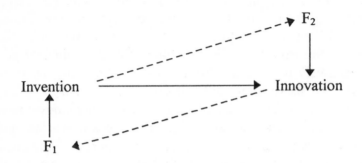

It is obvious that cultural differences, whether shaped by economic forces or not, can not only be observed between countries, but also within countries. Therefore, the question arises: what distinguishes managing laboratories at several locations within a country from multiple laboratories within different countries? This question is answered in the next section.

3.2 Higher variances between countries than within one country

3.2.1 Reasons

If it is optimal to run several laboratories for one company, then this may be the cause of specific benefits or costs that cannot be observed on a national level. If such influences would not exist, then the issue of internationalization would not be a specific one. Rather, all laboratories could be managed by the same model, regardless of their particular location. However, this does not seem to be so.

Using a number of case studies, De Meyer and Mizushima conclude: "We have no record of companies which would indicate that it is easier to do research and development on a global scale than in a geographically centralized approach. Globalization of research and development is typically accepted more with resignation than with pleasure. In one of our cases, internationalization was almost described as an unavoidable nightmare, closer to a marketing gimmick than to an effectively contributing research and development outlet" (De Meyer/Mizushima, 1989, p. 139). The growth of Swedish spin-offs is affected negatively by an increasing number of foreign subsidiaries charged with doing R&D work (Dahlstrand, 1997, p. 339). Why is it that these specific problems arise when crossing borders? We see the following reasons.

(1) By mentioning a gimmick the authors indicate that the terms "international", "foreign", or "global" may have positive meanings when it comes to attracting personnel or selling innovative products, even though identical objective criteria might be achieved by strictly national activities. This argument indicates that, in practice, laboratories may not exist for the rational grounds that were advanced in the preceding chapter and that will be developed further in later chapters. Here, we exclude such non-economic causes related to the existence of laboratories from further discussion.

(2) The differences in legal frameworks show more variation between countries than within one country, even if a particular nation is organized as a federation of states and is not governed by
one central authority. The legal framework of a region may support or constrain R&D in specific ways. Examples are found in:
labor law, intellectual property rights, security standards for
performing experiments, etc. Companies may cross borders because, in this way, they can avoid some constraints and make use
of legal regulations that support their objectives. Detailed
knowledge of the legal systems and their enforcement is a prerequisite for deliberate action.

At the same time, such differences may result in considerable
discomfort in international teams working jointly on one project.
This is observed in German firms which are required by law to
share proceeds from their empoyees' inventions with the inventors. The same is not required from their subsidiaries in the U.S.
Assuming a joint project at an American laboratory site with
German and American personnel as joint inventors, it would be
unacceptable to the U.S. researchers if the Germans received an
extra bonus. If the Americans were also given a bonus for their
inventions, their colleagues who were not working on international projects might ask for the same treatment. The resulting
problems are obvious.

(3) In most countries, we can observe that cultural variances are
greater between them than within any one of them. Some cultures are more supportive to a specific spectrum of R&D activities than others. For instance, the rejection of oral contraceptives
in Japan is a strong argument against placing relevant R&D in
this country. This reiterates the arguments shown in Figure 2
(above). Similarly, strong opposition to gene technology has
certainly influenced the decisions of German companies in the
chemical industry to seek opportunities for related R&D abroad.
Destroyed field plots of genetically manipulated plants cause extra costs and delays in the R&D which, in turn, reduce the expected sales from the eventually successful R&D. Thus, companies try to make use of the more favorable cultural environments
and to avoid the less favorable ones.

(4) Education between countries varies not only with respect to its duration or specific contents but also with respect to fundamental ideas that guide education. The basic ideas that shape the education of engineers in France, Germany or the U.S. are different (Lundgren, 1990). Such differences may lead to varying perceptions and approaches to problems as well as differences in solutions. It could be extremely helpful to make use of these differences, if the resulting cost and benefit variations would become known in detail. A large European automobile manufacturer remarked that its participation in European Union-sponsored projects was originally less motivated by the welcome hope of sharing a burden with others, but by the expectation to better understand these different engineering approaches and to eventually benefit from this experience.

Internationalized R&D might, therefore, contribute to the company's learning (De Meyer, 1989, p. 18). This might be further supported by international teams. If members of an international R&D team communicate effectively in the same language, and if they are top-level experts, they might be more creative than a strictly national team. The reason could be that problem solving is: "free from inhibitions. More tolerance is expected from people of other nationalities when unfamiliar opinions are voiced. Less immediate rejection of a new idea is expected on the basis that the reasons will need to be explained to the foreign national. This makes other members of the group think twice before rejecting an idea, and, on second thought, many ideas may be found acceptable" (Albach, 1993, S. 148). These views are based on cases, on interviews, and on experiences. Along these same lines, Honko argues that this tolerance of new ideas can be observed in many American R&D laboratories and that Finnish industry benefits from the fact that it is located in a small country that needs to integrate ideas, in particular from its Nordic neighbors and from Russia (Honko, 1990). Honko's arguments are based on casual observations and theoretical reasoning. Furthermore, teams that have members from a multitude of countries and laboratories within a firm are considered coordination mechanisms among laboratories (Gerpott, 1990, p. 243; Gerybadze/Reger, 1997, pp. 9, 10; Beckmann/Specht, 1997, p. 441).

However, hopes should not be raised too high. Interviews in German companies lead to the conclusion that one "cannot say that internationality is per se an important factor in determining a superior innovative team" (Albach, 1993, p. 149). To follow up on Albach's argumentation, it might also be true that the presence of foreign team members leads to the acceptance of too many weak ideas because of too little opposition to them. It could also be said that the interviewees have not been able to study the concept in a reliable way. Based on interviews in high-technology firms, it is observed that coordination of international teams is particularly difficult (Gerybadze/Reger, 1997, p. 11). Beckmann/Specht (1997, p. 441) and that such teams invite relatively more infighting as competences are put at stake. Thus, it might be a good idea for the company to hire novices for work on international teams and let them choose other occupations within the firm after project completion, rather than to invite "representatives" of different laboratories to become team members. Furthermore, it is difficult to argue that laboratory managers would agree to share their "best" personnel with other laboratories rather than to make use of this talent for their own research. Weighing the arguments, the issue definitely needs further research.

(5) Cultural differences, language problems, and real or imagined threats to foreigners might constrain cross-border mobility of specialized human work forces much more strongly than less explicit differences of the same kind within one country. To avoid these high costs of moving, a company may move its laboratory to where a scarce resource can be tapped.

(6) Cultural rivalries within international corporations may lead to setting up multiple R&D facilities. In problems involving high degrees of uncertainty, where parallel approaches to a R&D problem are called for, such strategy could well pay off. The same task could be attacked from quite different angles, and the laboratories - sparked by the additional cultural rivalries - could put additional effort into becoming the winner of a race for the solution. What appear to be cultural rivalries at first sight could also be competition for scarce resources. These are sought as they might establish a base of power. Power, in turn, might serve personal or organizational objectives. Interview results in leading

multinational companies indicate that a policy of charging one site with a world product mandate or selecting it as a competence center invites strong competition to be selected (Gerybadze, 1997, p. 28).

(7) Factor costs most often vary more between different countries than within one country. Relocation decisions in response to such variations might be ruled out for economic reasons. We do not know of a study on the economic implications of closing a laboratory site. The casual remarks of laboratory directors who have gone through this process indicate that it is lengthy and costly. This is particularly so in those countries where labor is strongly legally protected or where intensive co-determination legislation calls for negotiations on protective measures for all those who are threatened to lose their jobs and who are not willing to move to an alternative location, even though this could offer employment within the same company. Such inflexibility might be a reason for not considering certain countries as potential laboratory sites. Thus, instead of relocation, additional R&D capacity could well be placed so as to profit from these observed differences.

(8) Some countries restrict the economic activities of foreign companies, unless these companies secure a certain share of local content in their added value. This has been applied also to R&D. Hansen claims that these restrictions are applied with respect to pesticides in India, pharmaceuticals in France, or the exploration for and the production of oil and gas from the Norwegian shelf (Hansen, 1998). There are indications that the number of R&D people employed in Norway by foreign firms is correlated with the likelihood of gaining access to exploration and production activities. Such relationships lead to very difficult cost-benefit evaluations that go far beyond a few laboratories.

(9) Production of knowledge is neither distributed evenly over the globe, nor is it concentrated in only one location. The cost of accessing this knowledge may depend on its distance from the source. Distance is an operationalized construct that combines many types of influences from the psychological to the economical. In general, distances across a national border are much longer and much more varied than distances within one country.

The same applies to communication media that are employed to cover the distances.

(10) Customer needs and wants usually are much more diversified across nations than within anyone of them. This offers particular opportunities and difficulties at the same time.

3.2.2 Consequences

In conclusion, we see that the variance of potential benefits and costs of performing R&D in foreign countries can be substantially larger than the variance that arises from choosing the same number of locations within one country. In some instances, internationalization of R&D might lead to higher returns. Both effects offer threats and opportunities to management. Such management challenges are not mentioned very often. According to Medcof (1997, p. 316) in his literature review, few authors address specific management problems of internationalization, and those who do concentrate on cultural differences, the need to establish effective liaisons with other departments within the company, as well as the potential of modern communication technologies. A real challenge lies in finding a superior risk-return combination.

As shown in Table 3 below, some authors have concluded that making use of cost differences is a less powerful explanation of international dispersal of R&D units than is the exploitation of market factors or the uneven distribution of knowledge production. In regards to this point, De Meyer states: "Faster learning of more relevant information is . . . the key to explaining the internationalization of R&D. Learning about customer needs, monitoring the hot spots of the field to quickly learn about the most recent developments, and having access to the engineers and scientists who can process this information quickly is the objective of the internationalization process" (De Meyer, 1993). This technical learning might be supported by improving conditions of organizational learning, which means mastering the management of international R&D.

International management of technology uses three interdependent approaches to respond to the specific conditions outlined above (Gerpott, 1990, p. 228; Specht/Beckmann, 1996, p. 421).

– Firstly, management needs to select a specific organizational structure that assigns missions to laboratories and laboratories to locations. This structure has to determine as well which technologies should be developed internally and which should be developed cooperatively or acquired externally.
– Secondly, resource potentials need to be assessed and developed particularly in reference to human resources.
– Thirdly, processes of technology development need to be managed particularly in reference to those processes that cut across the borders of departments and - in doing so - also across country borders. Managing these processes involves the reporting, coordinating, and scheduling of tasks, as well as the supervision and development of joint visions, which - as we have learned - is more difficult between countries than within a country.

3.3 Benefits and costs of international R&D

Let us now have a look at the empirically established benefits and costs of internationally dispersed R&D facilities. Other authors use the terms pros and cons or centrifugal and centripedal forces (Beckmann, 1997, pp. 46 et seq.) synonymously. Interestingly, there are a considerable number of studies detailing benefits and costs, even though they are of substantially different scope and statistical quality. The main parts of the following tables (Table 1 and Table 2) are based on information originally compiled by Schmaul (1995). Studies at the industry level have not been included; all studies mentioned are based on data from individual firms. A few reasons for favoring or disfavoring internationalization of R&D, such as different R&D intensities at the headquarters and in foreign countries, diseconomies of scale in R&D, and different competences of R&D personnel, appeared only once in the studies. These reasons could not be represented in the tables.

The tables show the authors and the year when the studies were presented. It is then shown (type of study = Tp.) whether the study is a report of experiences as gained by practitioners (P), whether it is based on interviews (I), on written questionnaires (Q), or on a collection of a small number of cases (C) and their analysis. The countries involved (the symbols here appear to be self-explanatory, perhaps with the exceptions of CH for Switzerland, S for Sweden and G for Germany), the number of firms (n), and the number of foreign laboratories (N) are shown as well.

The reasons for foreign R&D laboratories are then grouped in three classes:

(A) Resource access, which comprises:
  recruiting R&D personnel at the foreign location (1),
  proximity to scientific competence centers (2),
  support for local production and customer services (3),
  use of low labor cost (4), and
  support for technology transfer (5).
(B) Product market access is another class that collects four more reasons, namely:
  proximity to customers and markets (6),

necessity to redesign products for local use (7),
climatic conditions (8), and
the presence of major competitors in the same market with a
similar engagement (9).

(C) A third class is composed of three more reasons:
political or environmental conditions that are more favorable
than at the headquarters location (10),
acquisition or chance events (11), and
the special initiative of local managers (12).

Most of the reasons given in Table 1 suggest that foreign R&D is often complementary to domestic R&D and only less frequently serves as its substitute. The latter can be expected, if domestic R&D costs exceed those in a foreign country or if political or legal restrictions are the major reasons for establishing R&D activities abroad. Also, powerful customers might require to have development resources close to their plants. However, for U.S. firms, it is concluded that: "there is little evidence . . . that much of the R&D undertaken abroad is meant to displace domestic R&D" (National Science Board, 1996, p. 4-44). Considering environmental influences, it has often been reported that the bio-technology research of German companies was placed abroad because of restrictions that this type of research faced in Germany. To the degree that this is true, it may not appear to be important, if the level of expenditure is related to all chemical and pharmaceutical R&D in Germany; however, the leverage that the activities exert cannot be overlooked.

The most frequently researched reasons show that foreign R&D laboratories are located where R&D personnel can be recruited easily or where the laboratories are close to scientific competence centers; another frequently mentioned set of reasons shows that laboratories are used to support local production or to provide customer services and are, thus, close to the market. These reasons imply quite different missions of the laboratories. While one group is more research oriented and eager to become part of the local scientific community, the other group is more application or development oriented. A third set of reasons relates to the use of favorable legal or political environments for R&D.

From this short overview, it is obvious that reasons interact with laboratory missions and that environmental influences cannot be neglected in decisions regarding location.

Table 1: Reasons for performing international R&D
(For explanations of the abbreviations, see the text.)

Author	Y	Tp	Countries	n	N	1	2	3	4	5	6	7	8	9	10	11	12
Papo	71	P	-	1	-	*			*			*	*				
Ronstadt	78	C	US	7	49	*	*	*		*	*						*
Behrman/ Mansfield/ Teece/ Romeo	79	Q	US	55	-						*						
Fischer	80	I	US, EU	49	206	*	*	*	*		*	*	*				
Pausenberger et al.	82	Q	G et al.	19	-	*	*	*	*	*		*	*		*	*	
Haug/Hood/ Young	83	I	GB	15	15	*		*	*		*	*					*
Håkanson/ Zander	86	C	S	4	-	*	*	*			*				*		
Harris	87	C	US,EU,J	8	-	*	*				*	*		*	*	*	
Freudenberg	88	I	US,EU	8	-	*	*		*			*			*	*	*
Perrino/ Tipping	89	I	US,EU,J	16	-	*	*				*	*			*		
De Meyer/ Mizushima	89	C	EU, J	22	-	*	*	*		*	*		*		*	*	
Krogh	90	P	-	1	-	*	*	*			*	*		*	*	*	
Zaininger	90	P	-	1	-	*	*	*			*	*					
Hakanson/ Nobel	90	Q	S	20	150	*	*	*	*		*	*			*	*	
Gerpott	91	I	US,EU,J	16	-	*	*				*	*			*	*	
Caluori/ Schip	91	I	CH	31	-	*	*	*	*						*		
Taggart	91	I	US,EU	22	-	*	*		*		*			*	*		
Pearce/ Singh	92	I	US,EU,J	128	-	*	*	*	*	*	*	*		*			
De Meyer	92	C	US,EU,J	14	-	*	*		*	*	*	*					
Håkanson; Håkanson/ Nobel	92 93	Q	S	20	151		*	*			*	*			*	*	
Serapio/ Dalton	93	Q	US	50	-	*	*	*			*	*		*	*		
Brockhoff/ v.Boehmer	93	Q	US,G,GB	22	59	*	*		*	*	*	*	*	*	*		*
Miller	94	Q	US,EU,J	20	-	*	*	*			*	*	*				
Papanasta- siou/Pearce	94	Q	J	-	-	*	*	*	*		*	*					
Beckmann/ Fischer	94	I	G	13	-	*	*	*	*		*	*			*		
von Boehmer	95	Q	US,GB,G	84	228	*	*	*	*		*	*		*	*	*	*
Gerybadze/ Meyer-Krah- mer/Reger	97	I	US,EU,J	21	-	*	*	*			*	*			*		
Kurokawa/ Iwata	97	Q	US,EU,J	-	250	*	*	*			*	*			*		
Total	-	-	-	-	-	26	25	19	14	6	24	22	6	7	18	10	5

A similar although much more restricted overview of empirical research can be composed of reasons raised against foreign R&D. These reasons are listed in Table 2. As before, authors, type of study, country, number of firms and of laboratories are shown in the first few columns. After that the factors inhibiting foreign R&D are shown, again excluding those that have been mentioned only very rarely. Two groups of reasons can be identified.

- In the first group, we include ten different resource-related reasons:
  the assumption of economies of scale (1),
  the necessity to secure a critical minimum size of a laboratory (2),
  difficulties that could arise with respect to securing a cohesive goal structure and coordination (3),
  more difficult communication (4),
  difficulties of protecting know how (5),
  increasing cost of foreign R&D (6),
  greater distance from the main locus of production activities (7),
  potential delays in projects (8),
  relatively less qualified personnel in foreign laboratories (9),
  little readiness of headquarters' R&D personnel to move to a foreign location (10).
- Let us now consider a second group of reasons. These market-related arguments are not mentioned quite as often as before. The arguments are:
  Foreign markets are too small to make R&D sustainable on this business (11), and
  a global orientation that is not supported by particular foreign laboratories (12).

Political motives remain unmentioned. Negative inducements of host governments, which include regulations on testing, government controls, or profit repatriation, can be disruptive to R&D activities (Behrman/Fischer, 1990, p. 109). Protectionist government policies for maintaining technological competitiveness at a regional or national level are manifold. These policies have been termed "techno-nationalism" (Stevens, 1990).

The assumption of economies of scale in R&D is the most frequently mentioned inhibitor of internationally dispersed R&D units. We already know that this assumption is not very strongly supported.

Table 2: Reasons against foreign R&D

Author	Y	Tp	Countries	n	N	1	2	3	4	5	6	7	8	9	10	11	12
Papo	71	P	-	1	-						*						
Behrman/ Fischer	80	I	US, EU	49	206	*	*				*			*			
Pausenberge et al.	82	Q	G et al.	19	-	*	*	*				*					
Harris	87	C	US,EU,J	8	-	*			*								
Behrman/ Fischer	90	I	US, EU	48	206	*	*			*	*			*	*	*	
Krogh	90	P	-	1	-					*							
Caluori/ Schips	91	I	CH	31	-	*	*	*	*	*		*	*				
Pearce/ Singh	92	I	US,EU,J	128	-	*			*	*					*		
De Meyer	92	C	US,EU,J	14	-	*	*	*	*				*				
Brockhoff/ v.Boehmer	93	Q	US,G,GB	22	59											*	
Casson/ Singh	93	Q	GB	21	-	*			*	*				*			
Miller	94	Q	US,EU,J	20	-			*				*	*			*	*
Beckmann/ Fischer	94	I	G	13	-	*	*	*			*						
Kurokawa/ Iwata	97	Q	US,EU,J	-	250				*	*	*			*		*	
Total	-	-	-	-	-	9	6	5	6	6	5	3	3	4	2	4	1

The same is true with respect to the assumption of a minimum efficient size of a laboratory. Arguments relating to communication costs and know-how protection, the latter involving costs of maintaining secrecy, are mentioned so frequently that their importance cannot be easily denied.

It is not too revealing to study these lists of reasons in much greater detail. They reflect the beliefs or perceptions of the managers. Some of these beliefs actually contradict each other and, therefore, cannot be generalized. This is not astonishing. Even companies with a comparable strategic intent might face environmental conditions for laboratory work that vary substantially between countries. The discussion of cost-related reasons, market-related reasons or even political stability by Behrman/Fischer (1980, pp. 16 et seq.) considers these reasons and alternatively makes this variation clear as either an obstacle to internationalization or a supportive argument. None of the studies presented the full profile of reasons to the interviewees which means that those reasons not presented had a smaller chance of being mentioned. This introduces biases into the overall picture.

Almost each foreign laboratory will exist not only for one, but for many reasons, and its existence might at some time be threatened for many reasons. Not all of the different reasons are independent of each other. Therefore, some authors have tried to aggregate the interdependent reasons into a few independent factors by means of factor analysis.

Based on importance evaluations elicited from U.S.- based laboratories of foreign firms, it is concluded that "both market- and technology-oriented activities are important, but that technology-oriented activities are, on balance, more important" (Florida, 1997, p. 90). However, this observation is not based on stable ground. The "activities" were not analyzed for potential interdependences, and the conclusions which were drawn result from weighing the three levels of the answering scale used for each of the activities according to the beliefs of the researcher. The degree to which such procedural details contribute to shaping the results cannot be evaluated. However, such results need to be treated with caution.

The most encompassing study to-date of this kind of investigation aggregates 38 different reasons for foreign R&D laboratories into eight factors (von Boehmer, 1995, p. 92) and rates these by their relative importance (Table 3). At first, it is interesting to note that the importance weights are substantially different from the frequencies with which the various reasons are mentioned in the studies shown in Tables 1 and 2. In particular, the attractiveness of the scientific environment in a host country, which is mentioned very frequently in the studies quoted in Table 1, takes only a median position in Table 3. This difference is supported by a study based on patent data which demonstrates that: "there is no systematic relationship between a country's sectors of technological advantage and the relative presence of foreign firms in those sectors" (Patel, 1996, p. 42).

− Firstly, this could mean that the cost of entry into such fields of technology is high for foreign firms, because they lack either their own R&D that would be necessary to interpret and to absorb the relevant knowledge or to offer some new knowledge in return; or the foreign scientific establishments have set up barriers of entry to non-local firms.

Table 3: Factors explaining foreign R&D activities and their relative importance

Factor	Importance weight
Availability of personnel	3.02
Market attractiveness	2.37
Management activity of foreign subsidiary	2.21
Market specifics	2.07
Attractiveness of the scientific environment	1.77
More liberal legal or financial environment	1.39
Relative factor cost (labor)	1.37
Matching competitor's moves or proximity to suppliers	1.14

Source: von Boehmer (1995, p. 101).

– Secondly, it is found that the relative importance of the factors varies with the different missions of the laboratories. This is another reason for not generalizing the information found in Tables 1 and 2.

It is interesting that the factors relate very naturally to the four determinants of competitive advantage that are identified by Porter (1990) on the basis of extensive case research. These are:

(1) demand conditions which include the structure of market segments, anticipatory buyer needs, demand size, rate of growth etc.;
(2) factor conditions including human labor, physical resources, scientific and technical knowledge, infrastructure etc.;
(3) firm strategy and its contextual conditions, including rivalry;
(4) related and supporting industries which include competitive advantages in supplier industries.

The relationships with the factors mentioned in Table 3 are so obvious that they need not to be made more specific. The conclusion to be drawn from this observation is that firms use their international R&D strategy as a determinant of their competitiveness. Empirical evidence of differences among countries which are related to the levels of determinants of competitive advantage is available (Pearce/Singh, 1992, p. 151).

A very important finding is that different levels of factor costs are of little importance in explaining location decisions. There might be a number of reasons for this.

- Firstly, productivity is more important than factor cost per unit of time.
- Secondly, costs may be of less importance by orders of magnitude than market factors.

The little relevance of cost data in co-determining laboratory locations could signal substantial differences between internal views and the less differentiated public arguments. For instance, many observers claim that hourly wage differences exist between engineers employed in different countries. A case in point is an alleged difference between high German and low Japanese wages during the late 1980's. A more detailed analysis does not quite support this view (Maringer, 1990). Working with data from companies in the electronics industry, Maringer compares personnel costs in Germany with those in Japan. It is true that labor costs per hour for engineers in their first jobs in industry are substantially lower in Japan than in Germany. However, considering the differences in educational levels of engineers who have graduated, and thus the time span necessary until engineers in both countries are ready to solve substantial problems on their own, and furthermore considering cost increases due to seniority as well as lifetime hours worked, it is found that the present value of the costs per engineer is practically the same in both countries (assuming reasonable interest rates to calculate the present values). This demonstrates some of the difficulties and possible traps of taking comparisons lightly.

Little relevance of cost arguments might arise, if comparisons are limited to highly developed countries. However, a number of firms have established development facilities in Eastern European countries or in developing countries to gain access to very low cost, well-educated personnel (Perry/Sigurdson, 1997, pp. 348 et seq.). In such location decisions, the major considerations besides costs are: achieving a high-enough level of productivity, securing secrecy, and maintaining adequate levels of communication.

To sum up, we conclude that many benefits and costs have to be considered to explain why laboratories charged with a particular mission exist under specific circumstances within a country. The weight of these factors apparently has little to do with the frequency with which the variables are mentioned in empirical studies.

In very general terms, it can be argued that managers should weigh the benefits and costs of any foreign laboratory and decide on its location on this basis. Whether this is done on the basis of scoring rule models, financial analysis, or simple reasoning is not really known. In fact, such evaluations are extremely difficult to make as most of the criteria undergo substantial changes over time and have very far-reaching, even indirect consequences. A conceptual view of these influences is presented in Figure 3. Also, reliable measuring scales are not yet developed for many criteria. The evaluations are even more difficult because of the multitude of approaches to the internationalization of R&D that were mentioned in the first chapter.

Figure 3: A concept of the influences on the existence of foreign R&D laboratories

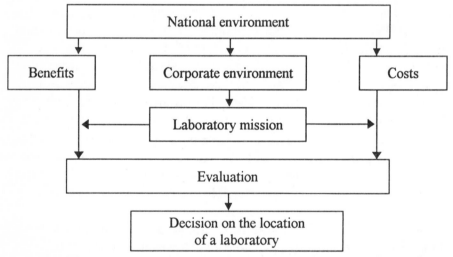

The view presented in Figure 3 reflects contextual influences on the weights of benefits or costs of laboratory locations. Qualitative results from an analysis of international R&D of selected German and foreign-owned companies operating in Germany indicate different importance weights for proximity of R&D to lead markets, national science systems or production by industry and type of R&D (Beise/ Belitz, 1998). In a more pragmatic approach the same idea has been used to structure R&D location decisions within a major company in the chemical industry.

4. Starting foreign R&D operations

4.1 Starting by cooperating on projects

As concluded in the previous chapter, a reasoning process supported by an evaluation of benefits and costs will precede the start of international R&D. In this chapter, we present various ways of starting up activities. In one of these approaches, the incidental acquisition of a laboratory, there might not have been a strategic or operational evaluation preceding the acquisition. Therefore, the question of how to integrate such laboratories into the technological strategy of a firm might be more difficult to answer than how to set up or acquire laboratories as a consequence of a technological strategy.

It is plausible to assume that some firms gain their first experiences with foreign R&D activities (excluding foreign employment in one national laboratory or the import of technological knowledge via buying patents or licences) by participating in a project that involves individual or institutional contributions from more than one country. Stevens states that: "Crossborder R&D cooperation agreements among semiconductor firms mushroomed . . . The arrangements offer an alternative to the legalities of joint ventures and allow work on new technologies which are not easily licensed" (Stevens, 1990, p. 44).

The European Union has been particularly active in supporting international, cooperative R&D projects. According to Teichert, technological, economic, and social-organizational motives determine the participation in such projects (Teichert, 1994, pp. 116 et seq.). Most of these motives are not centered on the project in question but extend far beyond it. A factor analysis of 13 items has helped to identify four motives. In summary, the following observations can be made:

(1) There exists a project-centered motive to enhance efficiency of operations. This factor is loaded with three items:
"Reduce or share uncertainty in R&D",
"reduce or share costs in R&D", and
"reduce time to market".
This motive relates to efficiency improvements, and it is clearly economical.

(2) All of the three other factors reach out beyond the project in question.

(a) The economical motive which relates to effectiveness enhancement. It is loaded with items that respond to the goals of "entering international markets" and "gaining market access by new products or expanding the product range". This shows a marketing strategic orientation.

(b) The technological motive shows an orientation towards specialization. It is loaded with the variables: "achieving technological synergies; developing complete systems", "concentrating one's own activities on core competencies", and "monitoring a wide range of opportunities". This supports the division of labor in bringing about technological change.

(c) The social-organizational motive reflects the effort to build technological networks. It is loaded with five items, namely "building an international R&D network for information exchange", "gaining access to partners' knowledge and technology", "observing and learning from partners' procedures", "creating an atmosphere of trust and familiarity with partners", and "learning how to cooperate for subsequent projects".

It is important to note the long-term perspective that exists in this last group of motivations. Economic motives and strategic motives of networking appear to have the greatest importance for companies entering into R&D cooperations. This has become clear from a conjoint measurement experiment that aimed to elicit partworth values of preferences for different types of cooperative arrangements. It was found that the highest absolute preference is achieving substantial cost advantages through the cooperation (Teichert, 1995, pp. 160 et seq.). In second place in the ranking order is the wish to cooperate with a non-competing partner. In third place is a variable measuring the intent to combine one's own activities with complementary resources or competencies. The strategic mission of building up new competencies and the preference for so-called pre-competitive research cooperations take the next higher ranking positions, however, at much lower partworth levels. This supports the results mentioned above.

Assuming four differently weighted motives for each of several partners in one cooperation, and a separate set of motives held by a sponsoring agency, R&D cooperations are constantly threatened by lack of a unifying vision. Their survival is characterized by a constant search for common goals and for compromises. The transaction costs that arise from these processes might easily be higher than the cost-savings or the benefits resulting from cooperative R&D. This is of particular relevance, if the sponsoring agency puts pressure on partners to accept other partners who would not have been chosen voluntarily. These are not the most likely partners for exchanging immediately marketable information. Research tells us that under such conditions the outcome of R&D cooperations is rated very low (Rotering, 1990, pp. 174 et seq.)

Strong, external competitive pressure might align objectives and goals of partners, and thus reduce transaction costs. This is very convincingly demonstrated in the historical analysis of the U.S. Microelectronics and Computer Technology Corporation (MCC) (Gibson/Rogers, 1994). This "consortium of consortia made up of 100 member companies with 340 full-time researchers" (Gibson/Rogers, 1994, p. 2) was the first for-profit research institution of its kind in the U.S. The Japanese threat or "the fear of not surviving the global economy has been an important motivator for public/private collaboration, however, such motivation does not sustain the co-operative/synergistic activity required to make such alliances work effectively over the long term . . . In marketplace economies, such collaboration . . . must be learned and reinforced" (Gibson/Rogers, 1994, p. 543). The heterogeneity of objectives for a cooperation can be expected to grow, if its activities become more market-related and if partners compete in the same market.

Another reason for strongly diverging motives might relate to the technological motives. In Europe, due to quite different traditions in engineering and science education, approaches to a task as well as a priority setting might differ considerably. Building a cooperation under these circumstances leads to transaction costs that can be expected to be high, even in face of dramatic competitive pressure from larger and more powerful markets or companies. Therefore, R&D cooperations might not be the first choice of the participating companies in responding to such pressure. However, it may be the only choice in a long term perspective because it initiates cross-cultural learning experiences that precede collaboration. The public support for cooperative R&D projects could be considered support for such a learning exercise that is ultimately aimed at reducing the transaction costs that arise from the national differences of relatively small countries. At the same time, subsidization might reduce the pressure on participating firms to find a common strategic alignment of their goals. It also gives motives of cost reduction a more prominent place than would be warranted without the public support.

In the presence of competitors, companies cannot rely entirely on cooperative R&D in developing strategic, market-oriented R&D approaches to support their business. If they choose international R&D cooperations to start internationalization on a project-by-project basis, then, new projects or new partners will have to be identified, if internationalization should be continued after the first projects come to an end. If the cooperation is not tied to a particular project and thought to continue as long as it appears to be useful with respect to its objectives, support has to be generated repeatedly to balance diverging interests.

4.2 Starting by establishing an R&D unit

In this chapter we do not asssume that R&D cooperation necessarily precedes an R&D unit. In Figure 4 we present major alternatives that could be chosen in order to establish an R&D unit.

Figure 4: A schematic view of establishing an R&D unit

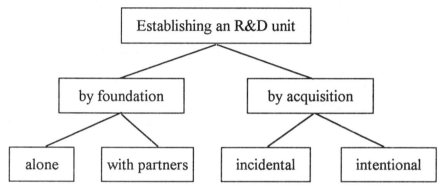

4.2.1 Acquiring laboratories

(1) The first of two alternatives is to acquire a laboratory (Chiesa, 1996, p. 9). This could be intentional, if the laboratory is the only or the major objective of the acquisition. This could involve additional assets to be acquired at the same time. Alternatively, an acquisition might be incidental. This occurs if an acquisition of some other assets cannot be completed without accepting the laboratory as part of the deal.

The relative frequency of acquisitions is mentioned in a large number of studies. Ronstadt (1978) reports that 26% of foreign R&D establishments had been acquired and that all of these were "incidental". More specifically, he states that: "none of the units was an 'intentional' acquisition, where an organization would be acquired . . . in order to gain access to the organization's R&D resources" (Ronstadt, 1978, p.8). Pausenberger (1982) found that at the end of the 1970s about 30% of the foreign R&D units of 19 mostly German companies resulted from incidental takeovers: the aim was not to acquire the R&D facilities. Behrman and Fischer (1990, p.25) show that 26% of the totally owned laboratories they studied were acquisitions. This figure is comparatively low. The reason is that the authors have excluded technical service units which, in part, had been acquired. Including these units in the sample should have raised the share of acquired laboratories. In an international sample of laboratories it was found that 20% had been acquired, though acquisitions were not quite as frequent in Great Britain as in other countries (Pearce/Singh, 1991, p. 193). In a study of U.S., British, and German headquarters and their foreign R&D units, 78 out of 227 cases (34.4%) had been acquired (von Boehmer, 1995, p. 64). Data from German companies in the chemical and pharmaceutical industry indicate that out of 30 laboratories 60% had been acquired. In one third of all cases, technological motives were dominant (Beckmann/Fischer, 1994, p. 639). These acquisitions might be called intentional. Later, Beckmann (1997, p. 1930) reports that 52% of the laboratories in the same industry are acquisitions. A slightly different view is presented by Håkanson and Nobel (1990), who report on both intentional and incidental acquisitions of laboratories. These authors (1993, p. 377) report that 61% of the laboratories of Swedish multinational companies were acquired.

Only 1.4% of the R&D conducting subsidiaries of Japanese companies were acquired; even after excluding the 15% of the cooperating units from the sample, the share of acquired R&D units merely reaches 1.6% (Yoshihara/Iwata, 1997). This percentage might indicate that the Japanese headquarters define a completely different mission for their laboratories as compared with the "Western" headquarters and their subsidiaries. However, the reasons reported in the study by Kuwahara and Iwata (1997, Table 1), who use the same data base, do not lend support to this assumption. It is, therefore, concluded that Japanese firms have an as yet unexplained preference for establishing their own foreign R&D, which is different from that of Western companies.

Strengthening a technological base is an important motive of mergers and acquisitions, although it ranks only in the middle positions of various strategic considerations (Chakrabarti/

Burton, 1983; Süverkrüp, 1992). This motive was studied with particular reference to cross-border mergers between Germany and the U.S. (Chakrabarti/Hauschildt/Süverkrüp, 1994). Of 325 cases that could be identified in the decade between 1978 and 1987, 26.5% responded. Of these, 28.2% were identified as "technology acquirers" or candidates for "intentional acquisitions". These are firms that want to have access to new technology and know-how. They strive for integration with the scientific environment of the acquired firm, or they are interested in the capabilities of the R&D personnel which they hope to enlist via the acquisition. Relatively more U.S. (62.5%) than German firms (37.5%) reveal this strategy, although the former are much smaller than the latter. In the majority of the cases, the acquiring and the acquired firms are within the same industry (79.2%). More than 70% of the cases studied are driven by other acquisitional motives. These include the incidental laboratory acquisitions. The figures suggest that these are more frequent among U.S. acquiring firms and less frequent among German acquiring firms.

It is less likely that laboratories with a research mission are acquired. Usually they are founded by the headquarters. This assumption is supported by statistical data from an international sample of firms (Pearce/Singh, 1991, p. 193). Generally speaking, some plausible relations between the mode of establishing R&D laboratories and their missions may be observed (Håkanson, 1992, p. 101). The history of the build-up of research laboratories might be different from that of development laboratories (Beckmann, 1997, p. 193).

To summarize these studies, there appears to be a trend towards absolutely more acquisitions of laboratories. This can be explained in part by the general observation of increasing merger and acquisition activities during the past decades. Also, relatively more foreign laboratories were established recently than in the past. At the same time, there are some national differences as well. It is not clear whether these differences reflect cultural traits or significantly different missions assigned to foreign laboratories. The latter explanation could mean that foreign laboratories are used as substitutes for particular knowledge or abilities that are lacking at the headquarters.

(2) Laboratory acquisitions pose particular managerial problems. The major managerial problem arising from incidental acquisitions is finding out how the laboratory might be used to support the competitive position of its new owner. At Nestlé, the acquired facilities are used for determining their strategic mission within the worldwide R&D portfolio of this Swiss multinational (De Meyer, 1988). The determination of a laboratory mission under such circumstances might require substantial cost and time. The newly acquired laboratory might not have immediate access to all information necessary for defining its mission. Help from other laboratories may not be available since the new laboratory might be considered a new competitor for scarce R&D resources within the firm. Therefore, threatening to terminate its operation may not be the last alternative because it is expensive as well.

(3) The first major managerial problem of intentional acquisitions is identifying suitable candidates. A substantial difficulty arises from the particular characteristics of knowledge which induce information asymmetries. These might be exploited by a seller who claims to know more than he actually does. Careful analyses of this situation is mandatory. At W. R. Grace, a special group within its Japan Research Center is charged with the mission of identifying candidates. This activity is called reverse technology transfer or licensing-in. Similar attempts are reported from other companies as well. It is obvious that such activities are costly. During nine months in 1996, Schering AG's technology office in the U.S.A. evaluated 147 innovative technologies, discussed potential acquisitions with 39 sponsors, started 6 negotiations, and completed three of them successfully.

One of the tasks of such units is to determine whether the mission of the intentionally acquired laboratories should be complementary or substitutive to the already existing laboratories. The trend in the early 1990's in European and U.S. companies of scaling down research activities particularly at headquartes (Brockhoff, 1997) sometimes coincides with an expansion of more market-oriented development in other countries by new laboratories (Gerybadze, 1997, p. 20). In regards to this point, starting or expanding dispersed R&D units not only has a quantitative dimension but also a qualitative one in terms of changing the thrust or composition of activities.

To find out whether the newly acquired laboratory could copy or broaden the already existing R&D activities one can compare fields of activities. To substantiate this, content analysis of reports or publications from the laboratories could be a first approach. Most simply, one may look for technologically characteristic keywords and the degree to which these are identical or non-identical in the documents. This approach is limited to publications on basic or applied research for two reasons. Relevant technological information describing development work is hardly ever published, in order to secure competitive advances. Publication of patentable work cannot be expected to appear prior to filing for patents because this would preclude future patenting.

Therefore, analyzing patenting behavior is an interesting alternative. To analyze the overlap of the fields of activity, a number of indices can be calculated. Careful analysis of the available literature has led Pieper and Vitt to suggest two scales that need to be considered in evaluating the technological relatedness of firms: one scale measures the degree of overlap among the technologies that two or more firms master, and the other scale measures the degree of transferability of such technologies. This might be rather low if these technologies build substantially on tacit knowledge (Pieper/Vitt, 1998). In intentional acquisitions, one looks for laboratories with small overlap. If transferability of technological knowledge is high, it might be used throughout the new parent organization. If transferability is low, either the newly acquired laboratory stays separate from the other laboratories of the firm, or substantial costs of transfer are encountered in bringing the acquired potential to fruitation in other parts of the organization.

Pieper and Vitt (1998) discuss various measures of technological overlap as they appear in the literature, and they develop their own measure. As shown in an empirical comparison, it has a number of advantages over other measures. This measure of overlap is defined as follows:

Let $n_{i,k,t}$ be the number of patents that the acquired company i owns (or, alternatively, has applied for, which is reported by the European Patent Office) in patent class k in year t. The total number of patents of company i in year t is simply found by summing over all k = 1, 2, ..., K classes. This is called $n_{i,t}$. To smooth out abrupt movements in patenting activities, it is suggested that one considers these activities over years t = -3 to t = -1. This leads to the following patent shares:

$$f_{i,k} = \sum_{t=-3}^{-1} \frac{n_{i,k,t}}{n_{i,t}}, \text{ all k.}$$

Let us now introduce the acquiring firm j. For the two firms i and j, it is suggested that one selects all patent classes in which both firms file patents during the relevant years. Thus, if

$$\sum_{t=-3}^{-1} n_{i,k*,t} > 0 \text{ and } \sum_{t=-3}^{-1} n_{j,k*,t} > 0,$$

an index c is set to represent this particular class k*. These patent classes are called c = 1,2,...,C, C ≤ K. Then, technological overlap is measured simply by:

$$OV_{i,j} = \sum_{c=1}^{C} f_{i,c}.$$

An advantage of this formula over Jaffe's approach (1989) is that it gives more meaningful results in the case of firms with substantially different sizes. In most acquisitions, this is of particular relevance, since the acquired firm is smaller than the acquiring firm.

A further alternative for identifying potentially interesting prospects for acquisition is to develop patenting portfolios (Ernst, 1996). An example is shown in Figure 5. The two-dimensional drawing indicates the relative patent position (the number of patents or patent applications per firm over the number of patents or patent applications of the firm with the most patents in the analysis) scaled on the abscissa between zero and one. On the ordinate, the relative growth rate of patents is shown. This is considered as a measure of attractiveness. All measures refer to technology fields (labelled TF1 through TF5) which are defined for the industry in question. Expert opinion is needed to define these fields in a manner specific to industry. The size of the circles represents the share of patents (or applications) per company per technology field over the total number of patents (or

Figure 5: Patent portfolio for firms A and B, and technology fields TF1 to TF5

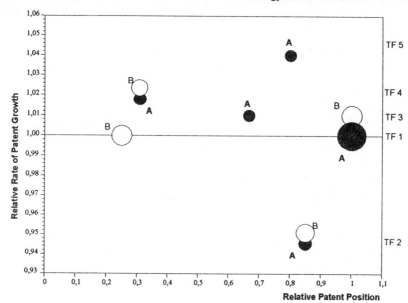

applications) for the same firm. To ensure simplicity in the illustration, we consider only two firms, labelled A and B. One firm (A) is the potential acquirer of another firm (B).

Company A is active in the same technology fields as company B (TF1 to TF4), and in addition, it holds patents in a fifth field (TF5). In TF5 there is another major patent holder which explains why A's relative patent position is less than one. In two fields (TF2 and TF4), A has a relative patent position that is identical to B. However, as indicated by the size of the circles representing these fields, the share of patents per company is smaller for A than for B. Thus, B has given these two fields more weight. Accessing this knowledge might be attractive to A in order to build a stronger competitive position. In fields TF1 and TF3, the relative patent position and the shares of patents of both competitors are just the opposite. This opens up an interesting opportunitiy to strengthen A's position, if it could acquire B. The opposite could be true as well. However, B being the smaller of the two competitors had obvious difficulties in acquiring A. Thus, as A saw the opportunity, it acquired B. The acquisition has substantially strengthened A's position by shifting all its relative patent positions to the right (except of that in TF5 where it does not acquire knowledge from B and that in TF1 where it already is the strongest patent holder).

(4) After identifying likely targets for intentional acquisitions, it becomes important to retain the relevant knowledge through the possibly arising frictions of the acquisition and post-acquisition phases. A difficult management problem will be to convince creative members of the acquired laboratory to become loyal to the new owner. This is expected for mainly two reasons.

- Firstly, the not-invented-here-syndrome may be present in both the acquiring and the acquired firm, particularly if these have been fierce competitors before. It would then be difficult to establish a joint culture in both groups.

- Secondly, acquisitions and mergers are mostly financed by additional credit capital, and this produces substantial financial stress to the new organization. It would not be uncommon if the new organization seek relief by reducing expenditures that bear fruit only in the more or less distant future. One of these is R&D expenditure.

The acquisition presents a new owner structure to the laboratory scientists and technicians. Only if this structure is valued positively, will it not destabilize their motivation. Otherwise, those facing more attractive employment alternatives might leave the company, and other laboratory scientists might reduce their creative efforts. Without specifically addressing international acquisitions, Vitt (1998) shows that the patenting activities of "major inventors" in acquired firms are significantly affected by the acquisition. Almost half of these inventors (47,5%) leave their laboratory within three years of the acquisition and only approximately half of them find other jobs within the same company. This high level of fluctuation should be compared with the level that would be registered in the absence of the acquisition. Such comparisons are hard to come by, but even informed estimates could put the figures in perspective. The remaining major inventors are found to reduce their patent applications by one third of the original level, while the quality of their patent applications is reduced by two thirds. Thus, the acquisition quite frequently has disruptive effects on the patenting activities of the original personnel. Certainly, this can be compensated by new personnel; however, additional costs are involved.

(5) Few analyses have been conducted on the success of intentional laboratory acquisitions. Representatives of "host-market companies" are quoted as seeing acquisitions as a fast track and easy route to internationalization (Behrman/Fischer, 1980, p. 33). However, as illustrated before, intentional laboratory acquisitions are not risk-free with respect to the acquisition of knowledge. While explicit knowledge that is well-documented can be traded easily, implicit knowledge that is part of human experiences cannot as easily be transferred from one organization to another.

Furthermore, success might depend on conditions pertaining to the technologies and laboratory cultures involved, as well as to the managing of the process of integrating the acquired laboratory into the absorbing organization. In one study (Süverkrüp, 1992), explanations for technical and economic success were identified on the basis of dyadic interviews in those firms where the motive of acquiring new technology was clearly spelled out and where both firms had R&D departments. From a path-analysis, we conclude that technical success is impaired if

(a) high technological uncertainty is present,

(b) if large cultural differences between the R&D units exist, and

(c) if the degree of formalization is high in the post-acquisition phase.

A great danger here is that companies which engage routinely in mergers and acquisitions, as well as those that notice large gaps between the production technologies favored by the firms involved, have a tendency to react to the apparent information asymmetries with more formalized approaches. This, as it occurs, is detrimental to success. With respect to economic success, we find that conflicts between technological "philosophies" have a strong negative effect. This supports another finding whereby a technological mismatch can derail expected economic success (Chakrabarti/Souder, 1984). Another negative effect is exhibited by the buyer's acquisition experience which indicates that there is little learning in these activities. Change of management teams in the pre- and post-acquisition phases can be another source of problems. As indicated by Haspelagh and Jemison, "many operating executives who will manage the post-merger integration are not included in the analytical process" (1987, p. 56). Clear objectives and support from experts strongly contribute to economic success. Summing up, we find that contrary to some views, it has to be concluded that the acquisition of R&D laboratories is not a risk-free approach (Specht/Beckmann, 1997, p. 439).

The present state of knowledge on intentional laboratory acquisitions and their success is meager. It is particularly disturbing that "indirect effects", like the introduction of more formalized processes or switching executives during different phases of a merger or acquisition, seem to have such strong effects. Not that these effects were not to be expected; rather, they call for a longitudinal research methodology because the effects occur with substantial time lags. This methodology has not yet been applied to the cases of interest. Furthermore, in studies performed to date, retrospective information had to be used in evaluating dependent and independent variables of interest. This might have introduced biases in the responses.

4.2.2 Establishing laboratories

Returning to Figure 4, we see that an alternative to acquiring laboratories is establishing them by foundation. Whether this is received by existing laboratories with praise or criticism will depend on missions and organizational relationships. A new laboratory over which an existing laboratory has no power and which has a mission to conduct rival research is probably faced with opposition from existing laboratories. Perhaps, however, they receive a warm welcome from hitherto badly served customer departments. Thus, defining mission statements for new laboratories is a very important task.

The selection of a mission for a laboratory is not entirely unbounded. Bounds might be established by the historically established structure of existing laboratories, by the degree of dependence on production or marketing units with which the laboratory is supposed to collaborate in a foreign country, and by the degree of independence of the founding organization from other partners. A partner in a joint-venture type of cooperation might be a university, a private research and development institution, a supplier, a customer or a competitor. It is reported that this type of R&D cooperation has grown (Hagedoorn, 1995; Duysters/Hagedoorn, 1996). In the 1970s, joint ventures and research corporations dominated cooperative R&D arrangements. These involve sharing equity investments between two or more firms. In the second half of the 1980s and the early 1990s, nonequity cooperative R&D agreements became the most frequently chosen form of partnerships.

While partnering may reduce the cost burden, it also means that eventual results have to be shared and that some of the strategic intent that follows the establishment of foreign R&D operations leaks out to the partners. This may be wanted, if the partners are suppliers or customers; it may be unwanted, if the partners are competitors or "neutrals", such as universities. Furthermore, the division of labor in a joint R&D venture can be a sensitive issue, particularly since the so-called "tacit" knowledge that arises in deriving the R&D results might be transferred back through the personnel in the partner's new laboratory to their respective home operations. The diversity of cultural backgrounds that enter a cross-border cooperative R&D arrangement tends to raise the transaction costs involved over those that might be expected, if a similar arrangement were sought at the national level. This may explain the observation that, until the late 1980s, the growth of such international cooperations in R&D was smaller than could have been expected from the frequent mentioning of alliances or similar arrangements in the management literature (Duysters/Hagedoorn, 1996).

Establishing a foreign laboratory without any partners reduces some of the difficulties mentioned above. A detailed discussion of organizational structures that describe relationships between laboratories involves many issues. This is why it is postponed to later chapters (Chapters 5 and 6).

4.3 Process steps

It is not known whether companies follow a strategy in choosing between the alternatives of setting up foreign laboratories or whether they take ad hoc decisions. Data from Finland seem to indicate that companies have developed preferences for either establishing or acquiring foreign laboratories (Hölsä, 1994). Interestingly, the "acquirers" of foreign laboratories spend relatively more on foreign R&D than the "establishers". Whether this reflects special inefficiency, differences of task structures or size effects is yet unknown.

The complicated and complex considerations on alternatives for establishing international R&D units should be assigned to a procedure, which should ensure that important issues are not overlooked.

One of these procedures was developed by Booz, Allen and Hamilton on the basis of consulting experiences and interviews (Gerpott, 1990, pp. 233 et seq.).

– Firstly, Booz et al. suggest analyzing the technological position vis-à-vis competitors and the future attractiveness of technologies. If both are high, a laboratory should be established either from scratch or by shifting resources to a new location. If the attractiveness is high, but the competitive position low, it is suggested looking out for acquisitions or exploring joint venture opportunities. All other cases could be treated nationally.

– Secondly, the frequency and importance of a new laboratory's communication with customers, internal departments or the scientific community should be used as a criteria for choosing a location. The maturity of a technology and the originators of creative inputs are considered focal in determining frequency and importance of communication.

– In a final step, the existing structures, the overall locationing strategy as well as external constraints should be integrated into a choice of a laboratory site.

Earlier criticism of this concept refers to a lack of operationalization of the decisive factors and of specific methods of their integration; it is noted that the concept is technology oriented. This could lead us to overlook the fact that products and processes tend to integrate a vast variety of technologies. All of them have to be available to potential new production and marketing activities in a foreign country. The question which arises is how this can be secured? This becomes particularly relevant, if the technologies have various degrees of maturity (Specht/Beckmann, 1997, p. 424). As a consequence, another concept is suggested (Specht/Beckmann, 1997, pp. 424 et seq.).

The second concept suggests improving on any given situation by the employment of two "levers", namely quality of interaction among all relevant institutions inside and outside of the company, and quality of the resources employed in R&D. In addition, the costs and risks involved in the structural change should be assessed. This information might serve as feedback, thus establishing a cybernetic model of process evolvement. In further detail, eight steps are suggested:

(1) Delimination of the field of action. Most often it involves only a small part of an organization. However, the choice limits potential, overall improvements or feedback with other functional areas.

(2) Selection of strategic requirements. This serves to separate activities within the organization from those that might be performed externally. However, the pro-active definition of core technologies certainly is not easy and is open to debate.

(3) Description of the present situation by certain criteria. These describe missions, sizes, interdependencies, autonomy, potentials etc. of existing units. Most relevant would be an assessment of competencies. To secure unbiasedness, outside help might be needed in such an audit.

(4) "Creative combination" of alternative missions (research, development with worldwide or regional mandate, application engineering) should be applied to generate alternatives. This appears to be a weak point, as it limits the approach substantially and is not explicitly related to the levers mentioned before. These appear only in the next step.

(5) Evaluation of R&D core processes of the alternatives with respect to necessities of interaction and barriers of interaction. These include barriers of communication. This step should lead to a pairwise comparison of interaction quality.

(6) Evaluation of the resource potential, including resource costs, productivity differences of R&D personnel, and possible effects of taxation. Here, the concept of optimal laboratory size with re-

spect to costs comes into play. Also, the potential threats from political or economic influences should be evaluated. Although this step involves the most complex issues, little methodological guidance is provided.

(7) Evaluation of the costs of implementation. Here, the authors suggest ranking costs, duration, and risks of the alternative processes of changing the organization and the missions. To what degree their own evaluations are valid cannot be established.

(8) Integration and choice. Again, many evaluation problems and a lack of a systematic aggregation of the evaluations of the three basic criteria are criticized (Specht/Beckmann, 1997, p. 437).

While some plausible suggestions are presented, the lack of an overall and well-tested procedure is all too visible. Also, to make choices, a close look at organizational alternatives and their likely consequences is important. These will be developed in the next chapters.

5. Task assignments to foreign R&D laboratories

5.1 Static classifications

In a recent study of industrial research laboratories, it was found that they need to meet four necessary conditions to be successful (Brockhoff, 1997). Adopting a resource-based view, it is suggested that the conditions result from building four potentials, namely:

(1) a potential to identify possibly relevant technological knowledge outside of the company in question,

(2) a second potential to absorb whatever knowledge has been identified as relevant and to make use of it,

(3) a third potential to develop creatively new ideas on the basis of the available knowledge (which is a primary source of unique competitive advantages), and

(4) finally an internal transfer potential. The last potential allows the new knowledge to become known and accessible to departments further down in the value chain.

If a firm establishes many laboratories, it becomes feasible to let these laboratories specialize on some of these potentials, particularly if this seems to offer greater advantages than assigning all these activities to each one of the laboratories. However, this assignment of potentials should neither apply to the outreaching nor to the inward communication functions. Thus, specialization is limited. However, within the intelligence, the absorptive or the creative functions there could be further specialization.

The idea of specialization is well-represented in a recent study of 238 foreign R&D sites of 32 leading multinational companies (Kuemmerle, 1997). It is found that: "45% of all laboratory sites were home-base-augmenting sites, and 55% were home-base-exploiting sites", in which the former type "is established in order to tap knowledge from competitors and universities around the globe; in that type of site, information flows from the foreign laboratory to the central lab at home. The second type of site - home-base-exploiting site - is established to support manufacturing facilities in foreign countries or to adapt standard products to the demand there; in that type of site, information flows to the foreign laboratory from the central lab at home" (Kuemmerle, 1997, p. 62).

It is obvious that the home-base-augmenting laboratories concentrate on developing the identification and the absorptive potentials, while the home-base-exploiting laboratories develop their strengths by building the internal transfer potential and, to some degree, the potential to develop creative ideas. The home base laboratory is implicitly charged with the broad mission of covering the build-up of all potentials (See Figure 6). Accepting this type of division of labor, one has to draw different conclusions for the two types of foreign R&D sites with respect to location decisions and the development of laboratory effectiveness. Let us assume that the communication frequency and distance determine location costs (Brockhoff, 1997). The location of the home-base-augmenting laboratories should then be close to where new knowledge is developed and can be accessed; this need for close proximity may be alleviated by fostering cooperation among institutions or individuals. However, the internal transfer capabilities need to be developed as well, in order to avoid that the results of the foreign laboratory remain unused. Temporary transfer of researchers, travel funds, interlocking memberships on decision making committees, etc. could be used to support this purpose and will be discussed later. However, even the architectural design of recently built laboratories is indicative of the experience that communication media cannot substitute for personal interactions. While the home-base-augmenting laboratories are frequently mentioned, for instance as "listening posts" in company reports and in the literature, they are missing from some of the earlier taxonomies of foreign laboratory mandates.

The location of the home-base-exploiting laboratory should be close to manufacturing or marketing locations abroad. Its staff needs to interact less with the headquarters laboratory, but mostly with the units which the laboratory has to serve and their customers.

Figure 6: R&D functions and laboratory specialization

Identifying external new knowledge	Absorbing external new knowledge	Developing creatively new knowledge	Transferring new knowledge

Home-base augmenting laboratories	Home-base exploiting laboratories

The identification of only two laboratory missions by Kuemmerle (1997) - home-base exploiting and home-base augmenting - resumes a discussion started by Cordell (1973) who presented the idea of "support laboratories" and "international interdependent laboratories". It excludes what others have called multi-motive units (Håkanson/Nobel, 1990) or "allround R&D units with a global focus" (von Boehmer, 1995) which need to be attached to a headquartered home-base. This classification leaves little room for laboratories charged explicitly with performing the function of creative development of new knowledge. Where this creative development of new knowledge becomes a primary function, a further type of laboratory can be identified. This has been called the "internationally interdependent laboratory" (Hood/Young, 1982; Pearce, 1989). It need not have a connection with other host country operations, but it is charged with a "world product mandate". Such a mission allocates complete and exclusive responsibility to a laboratory for a wide range of activities or for a particular product of the multinational enterprise. The laboratory plays the role of the competence center for a certain range of product-related activities. In an early phase, headquarters conceive, implement and coordinate the international R&D program to which this type of laboratory makes a distinctively specialized contribution. Once a laboratory has built a recognizable level of competence, it may well emancipate itself from the headquarters by exercising expert power. Whether its expertise then can be brought to fruition by integrating it with overall company objectives or whether this leads to clashes with all sorts of negative consequences is an open question. It is certainly a question that deserves substantial managerial attention. This attention is necessary because a product mandate usually covers several technology fields, which might be

relevant for other product lines as well. This could lead to an unwanted duplication of efforts. An alternative concept is that of a
"world technology mandate" that assigns responsibility for a particular technology to one laboratory, regardless of where the product and
process customers for the technology are located. Locally available
competencies, the resulting interfaces from any such selection of
models, and the distance that their inward and outward communication has to bridge are among the important criteria to be evaluated in
preparing an organizational choice.

Home-base independent knowledge, thus, could be a third type. It
would present radically new technology to create a certain product or
process or to diversify into something completely new. To generate
this type of new knowledge, any economically feasible location might
be chosen without regard to existing in-house laboratories.

In fact, recent empirically motivated research has identified up to
five different mission orientations for foreign laboratories (von Boehmer, 1995, p. 87; Håkanson/Nobel, 1993; Pearce/Singh, 1991, pp.
194 et seq.). Even earlier research had identified more missions: for
instance, Ronstadt (1977) considered it important to differentiate
between "Technology Transfer Units" and "Indigenous Technology
Units", which both serve local customers. While the former only
provides technical services on the original products, the latter devel-
ops differentiated products to serve the special needs of foreign customers. While both make use of competencies developed at the
headquarters' laboratory, the level of creativity that is applied in these
units is different. Thus, home-base exploitation could occur in various forms.

Chiesa's classification (1996) builds on Ronstadt's arguments to a
substantial degree. Besides the support laboratories for current products and processes, he identifies exploitation laboratories and experimentation laboratories - actually, not a very precise denomina-
tion, as exploitation laboratories might engage in experimentation.
While the former extend the knowledge base by the current resources
in the short term, the latter explore new technologies with a long
term (more than three years) perspective. The author stresses, however, that this is not necessarily a function of a central laboratory,
although the managerial decisions for experimentation projects would
have to involve the company headquarters. As technology scanning is
considered a part of the experimentation laboratory's activity, we can

clearly see that the classification criteria used here is different from those used by Kuemmerle. For each of Chiesa's categories a further criteria, namely the degree of dispersion of external technological knowledge, is applied to define subclasses of laboratory types.

More differentiation of task assignments to laboratories signals that management problems are also more differentiated than would appear by only looking at the two categories mentioned at the start. However, it also reflects al lack of clarity in dealing with the concept of R&D proper, as was already mentioned in the first chapter.

In Figure 7, we try to summarize the major taxonomies of foreign R&D units. The figure reveals quite well that none of the taxonomies seems to have captured all of the mandates that could be assigned to foreign R&D laboratories. Only two of the studies mentioned are based on rather large-scale empirical research.

Håkanson/Nobel (1993) report on the relative number of laboratories with a certain mission as well as the share of their employment. In Figure 7, this is reflected by the two numbers in brackets in consecutive order. From these numbers, we can conclude that production support units and multi-motive units are relatively small, employing a smaller share of personnel than their share of units, while market-oriented units and politically-motivated units are relatively large. In the case of Swedish firms and before Sweden joined the European Union, this might have been caused by attempts to establish a strong foothold in this market, because of the fear that Sweden might decide not to join it.

In Finnish multinationals, 22% of the foreign laboratories are charged with support missions in the sense of Pearce (1989), 33% perform tasks for local units, and another 45% are "internationally independent" laboratories (Hölsä, 1994). Only if certain categories of the Håkanson/Nobel (1993) classification are merged, we arrive at comparable shares. This indicates a mix of missions in the categories used in the less deeply structured study. The Finnish study indicates an interdependence of reasons for international R&D and R&D functions with more market orientation being found in the more local units.

The relative number of laboratories is also reported in a study by von Boehmer (1995). The numbers in this study do not match with those given in the Swedish study. Even though laboratories appear in

Figure 7: A summary of taxonomies of foreign R&D units

Kuemmerle, 1997	Hood/Young, 1982; Pearce, 1989	Ronstadt, 1977	Håkanson/ Nobel, 1993	von Boehmer, 1995	Chiesa, 1996
Home-base-augmenting laboratory (45%)			Monitoring Research (9%/8%)	Allround R&D unit with local focus (47%)	
Home-base-exploiting laboratory (55%)	Support laboratories	Techno-logy trans-fer units	Production Support Units (14%/5%)	Local problem solver (19%)	Production support units; Support labora-tories
	Locally integrated laboratories	Indigenous technology units	Market oriented units (21%/32%)		
		Corporate technology units		Cooperating basic researcher (5%)	Exploitation laboratories Experimentation laboratories
	International interdepen-dent labora-tories	Global technology units		Applied re-searcher (20%)	
			Multi-motive units (37%/22%)	Allround R&D unit with global focus (9%)	
			Politically motivated units (19%/34%)		

comparable categories in both studies, this inconsistancy might indi-cate different strategies on the part of the firms involved in the two studies as well as different perceptions of the missions.

The examples of taxonomies shown in Figure 7 make it clear that the authors have chosen different criteria or dimensions of classifica-tion. In an attempt to "derive a single, generally accepted scheme", Medcof (1997, pp. 302 et seq.) identifies seven bases for classifica-tion:

1. The type of technical work being done (research, development, support).
2. The other organizational functions with which the laboratory should collaborate (marketing, manufacturing, both, none).
3. The geographical area over which collaborating units are spread (local, regional, international, referring to the location of all collaborating units).
4. The organizational locus of output utilization (as in (2), above).
5. The geographical locus of output utilization (as in (3), above).
6. The geographical locus of coordination.
7. The mode of establishment (greenfield foundation, incidental acquisition, intentional acquisition, conversion).

Combining all the possibilities of these criteria would lead to more than 5,000 laboratory types. Not all of these are feasible solutions. Therefore, the author considers the first three criteria as most important and uses them for his classification. Thus, the locus of decision making, that we consider to be a very important criteria, is not taken into account explicitly. Also, motives of establishment are not considered relevant. In theory, Medcof could identify 36 different types of laboratories by applying the three criteria. By excluding some combinations on grounds of implausibility, he arrives at only eight cases to be considered in more detail. The principle of plausibility might not always be a good guideline. Thus, the assumption that a laboratory charged with the mission to perform research does not need to collaborate with other functions might apply to pure basic research but hardly to more applied research than is relevant in firms. Another lever that serves to reduce the number of cases is a possible interdependency among criteria. Interdependencies might exist between a laboratory's main mission and its status as acquired or set up. Acquisition leads neither to laboratories supporting local problem solving in von Boehmer's sense (See Figure 7) nor to all-round development with a global focus. This is suggested by data collected from 12 firms in the German chemical and pharmaceutical industry (Beckmann/Fischer, 1994, pp. 637 et seq.).

The benefit of the new classification scheme for eight types of laboratories is that one may comprise many known existing schemes. These eight cases are:

1. Local research units (corresponding to Ronstadt's corporate technology units)
2. Local development units (corresponding to technology transfer units or local problem solvers as mentioned in Figure 7)
3. Local marketing support units (which correspond to market oriented units in Figure 7)
4. Local manufacturing support units (which have a relation to production support units)
5. International research units (similar to corporate technology units or cooperating basic researchers)
6. International development units (which correspond to global technology units or applied researchers)
7. International marketing support units
8. International manufacturing support units.

In practice, however, one company might operate each of the cases (1) through (4) or (5) through (8) parallel to each other, and one laboratory might have several missions such as (3) and (4).

While such taxonomies offer insights into potential missions or task structures, they do not address the questions of decision making and coordination. Furthermore, taxonomies appear to be static. It is an interesting question to ask what we know about the changes of missions or tasks.

5.2 Views on dynamics

The taxonomy of R&D units (Figure 7) should not be interpreted dynamically by assuming that it hints at a "natural" development from one of the mandates to another one. One such interpretation (Figure 8) is depicted, although not fully supported, by Gassmann and von Zedtwitz (1996, p. 9). Interestingly, the figure takes two starting points, either a "scanning unit" or a "technical center". The former is thought to develop into a "listening post" and from this into a small "research laboratory" and finally into a "research center". The latter adopts the task of "product services", "application development" and finally "new product development". It is not made explicit, whether or not the two branches merge after having developed into the last stage. Interestingly, manufacturing support does not play a role in this scenario.

Figure 8: Evolutionary view of R&D mandates

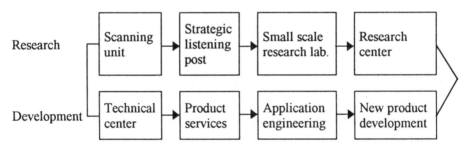

Source: Gassmann/von Zedwitz (1996, p. 9).

Earlier research (Ronstadt, 1977; Gerpott, 1990, p. 227) that tries to make such developments plausible was obviously strongly influenced by life-cycle concepts as they appeared in marketing and in the foreign trade literature. According to this concept, some countries were developers and producers of new high-technology products and other countries were the consumers or users. The latter could not be considered as likely laboratory locations. This concept has become less convincing during recent years when new products are released simultaneously around the world or where they are launched in different places according to a well-devised marketing plan within a short period of time. The sequence of countries for the presentation of a new product generation could well differ from the one chosen when the preceding generation of products was launched. Also, as a consequence of the increased globalization of competition, the evolutionary approach appears to have become outdated. Further, contextual conditions might change over time and might impact the decisions made by management. For instance, the locus of scientific activity might move from one country to another. Forced or voluntary "brain drain" is one of the reasons for this. When the most talented researchers leave laboratories in developing or Eastern countries to take positions in "Western" countries, firms are forced to redesign the portfolio of projects assigned to their laboratories.

Finally, it can be observed that organizational solutions are determined by a multitude of factors - many more than are represented in what might appear to be a natural enlargement of tasks. If the individual firm is chosen as the level of analysis, it may become extremely difficult to attribute reasons for internationalizing R&D as these can differ over time and with respect to each foreign location. This might neither be remembered in later interviews nor explicitly documented in a company's files.

Still, the strategic intent of a firm might have an influence on the formation of an evolutionary pattern of laboratory assignments. World-market companies "typically do not share the evolutionary pattern, from technical services to higher levels of responsibility, that was found among host-market companies" (Behrman/Fischer, 1980, p. 29). In spite of this argument, the authors continue to observe some sort of evolutionary pattern even in the newly created units for companies with a host-market orientation. This view brings strategic issues into play in determining a laboratory mission. However, strategic intent is another variable and not a constant. As the intent changes, so could the resulting patterns.

In addition to conceptual reasoning, it is difficult to assess any general evolvement of laboratory missions from empirical data. Given the small numbers and an undetermined time frame, it is only indicative of the fact that a minority of all laboratories (some 6%) found their mission curtailed, while some 39% could expand their mission (Beckmann/Fischer, 1994, p. 639). Beckmann (1997, p. 192) reports 16 mission changes among 35 laboratories in which the time frame is not clearly defined. Similarly, it is found that, on average, foreign

R&D sites of Swedish multinational firms have, over time, adopted a broader spectrum of responsibilities in the development of new technologies (Zander, 1997). The developments of missions can neither be related to the type of establishment nor to other influencing factors. Thus, the dynamics of mission development is an almost completely unexplored area, and it is certainly unexplained up until now.

Many more considerations for defining organizational structures, for instance Perlmutter's ethnocentric, polycentric and geocentric structures (Perlmutter, 1969), or ranges of activities, such as Levitt's (1983) or Porter's (1989) center for global, globally linked, locally leveraged and local for local concepts, have found applications in the field of internationalization of R&D. Interestingly, very little research has gone into attempts to relate the benefits and costs of internationalization to the definition and choice of such strategies.

In summary, we conclude that laboratory mandates can and will change over time. However, we find little support for the earlier idea that this process follows a regular, easily predictable path from one mandate to another. The large number of criteria that co-determine laboratory missions could all change in rather unpredictable ways. As a consequence of this, laboratory mandates might change as well.

5.3 Locus of coordination and locus of performance

5.3.1 A descriptive overview

There can be various ways to find a place for a foreign R&D unit on an organization chart. This is impressively demonstrated on the basis of case research which demonstrates many different solutions (Behrman/Fischer, 1990, pp. 125 et seq.). We shall not report on the various alternatives. Rather, we are interested in using optimum laboratory size and the necessity of coordinating R&D activities as guiding principles for generating alternatives.

On the one hand, we want to discuss a locus of decision making in the firm. If we assume that the firm operates a headquarters laboratory and possibly a number of foreign laboratories, then determining the mandates of these laboratories is of major importance. This encompasses a solution to the problem whether laboratories should be specialized such that each is responsible for performing a certain type of work alone or whether various laboratories should cooperate on a regular basis to carry out the tasks chosen by the decision makers.

In the classical hierarchical organization, decisions are made at the headquarters. For many years, researchers have known that due to the asymmetric distribution of information on business opportunities, it might be advantageous to make decisions at a decentralized level, such as profit centers, strategic business units, economically independent subsidiaries, joint ventures, etc. Some of these might be located in foreign countries to respond to the particular needs of the respective markets. Then, they might gather information relevant for defining profitable R&D tasks which they could use as a basis for their decisions. In between the two extremes of centralization and decentralization, good reasons for joint decision making between headquarters and, for example, foreign subsidiaries might arise. The reasons for this are found in eventually profitable knowledge spillovers to units other than the decision making unit, the avoidance of economically unreasonable parallel activities, or the necessitiy to jointly use some scarce resources. For these purposes, the development of shared objectives, pricing mechanisms for scarce resources, and standards for monitoring and control may all need to be jointly negotiated. On the other hand and as already briefly mentioned, task performance could be centralized at the headquarters; it could be decentralized in the sense of assigning tasks exclusively to one or more of the non-headquartered units, and a mixed mode could arise where headquarters and other units are involved jointly in problem solving. This is particularly relevant, if the optimum size of some laboratories has been reached and further expansion is less favorable than the establishment of a new unit. A combination of these two criteria leads to a 3x3 table of possible arrangements (See Figure 9). Not all of these cases may be attractive to a company.

In Figure 9, field (1) is characterized by the least overlap of tasks and by the least potential competition. Assuming internationally dispersed marketing activities, it is most likely the furthest away from day-to-day customer needs. Should decentralized laboratories exist, they are most likely to serve the headquarters laboratory by providing data and information on local conditions of product usage or on local scientific activities. Quite the contrary can be said for field (9). This case is best matched by exploitation R&D that is performed by an isolated specialization structure. It is chosen if there is a low dispersion of external sources of market knowledge. The R&D administration rests at the foreign unit's (or a business unit's) location.

Major problems might be that little long-term, strategic R&D work is performed, that resources might not be allocated optimally because of coordination problems, and that synergism cannot be developed systematically.

The remaining fields are characterized by attempts at balancing the benefits and drawbacks of the extreme cases in particular ways. For instance, fields (4) and (7) use locally available information to decide on a research and development portfolio of a central laboratory. Potential synergism in the production of new knowledge might be used. However, the distances between laboratory and decision makers might outweigh the potential benefits; the use of the scarce resource laboratory capacity might lead to conflicts; and the optimum laboratory size might soon be surpassed. In the case of field (7), a lack of strategic orientation adds to these problems. As the cases mentioned here are not of primary interest with respect to internationalization, we will not continue to discuss them.

Figure 9: Locus of decision making and locus of task performance

	Headquarters decisions	Mixed-mode decisions	Foreign site decisions
Head-quarters site per-formance	(1) Central R&D laboratory	(4) Joint decision making for central laboratory	(7) Local decisions, contracting R&D with central laboratory
Mixed-mode per-formance	(2) Division of labor among laboratories or cooperation of central and other laboratories	(5) Joint decision making and division of labor in task performance, including central laboratory	(8) Local decisions on task assignments to various central and decentral laboratories
Foreign site per-formance	(3) Task assignment to decentralized, foreign laboratories	(6) Joint decision making and division of labor in task performance, excluding central laboratory	(9) Local decisions and task performance in decentralized, foreign laboratories

While the taxonomy of Figure 9 is developed on the basis of two general criteria, an interesting question is whether at least some of the remaining and relevant fields can be identified among foreign R&D laboratories. Case studies by Chiesa (1996) indicate that most of the fields can correspond with reality, even though a lack of clarity in the descriptions remains. More broadly based studies by Schmaul (1995) support the existence of some of these types as well. These two sources will be used to characterize the situations.

Case (2) is characterized by a decision making unit at the head-quarters. However, the type of these decisions is very dependent on the particular tasks performed by the decentralized and the headquar-ters laboratories. Here we can see two different cases. Firstly, what Chiesa calls "supported specialization structure" of exploitation re-search is characterized by one center that has global R&D responsi-bility but is supported by scanning or adaptive units. Therefore, very detailed R&D planning and control is performed at the headquarters. Secondly, "experimental, specialized contributor laboratories" with complementary contributions from different units are assigned to this box. Here, an "integrator center" needs to coordinate and integrate the various contributions.

As argued with respect to case (1), information asymmetries might constrain the effectiveness of the approaches mentioned here, while the efficiency could be high. In earlier research on the internationalization of R&D, it was mentioned that laboratories could interact and be coordinated in three different ways. These were described as the hub, the competence center, and the network models. In the hub model, it is assumed that a central laboratory, preferably at the headquarters, serves as the decision-making unit and coordinates all other laboratories along the spokes that radiate out from it and represent the information exchanges. The foreign laboratories are primarily occupied with product adaptation to local needs or development work for the same purpose. Schmaul has found that 35% of the organizations he studied have chosen this model (Schmaul, 1995, p. 80). The hub model corresponds very well with the situation described in case (2). It is chosen when the number of international laboratories in a firm is relatively small.

In case (3), we expect to observe laboratories preoccupied with experimental R&D and isolated specialization. "This structure is chosen when there is an external world leading center of excellence . . . and a firm needs to be present to take part in the process of knowledge production and to attract key researchers . . . The prerequisite for choosing this structure is a low dispersion of the firm's technical activities in the particular field" (Chiesa, 1996, p. 16). As before, decisions are made at the headquarters, but not necessarily at a headquarters' laboratory. This may not exist or may not assume a status which is different from that of all other laboratories. It is very difficult to imagine the acquisition of knowledge for R&D decisions at a headquarters with this particular structure.

Case (5) is best described as an "integrated network model" of central and foreign R&D laboratories, characterized by tight controls, high subsidiary involvement in the formulation and implementation of strategies, and close ties among the different R&D sites (Håkanson/ Zander, 1988). It is also implied that companies which know how to manage this structure well are successful: "A firm's innovative potential lies in its ability to capitalize the resources of its various subsidiaries, to integrate the assets and the capabilities of the different units, and to leverage the unique strengths of resources of each unit to generate innovations to be exploited worldwide . . . Knowledge is developed jointly and shared worldwide. The central focus of the organization is to exploit the learning processes that take place at each subsidiary, and integrate and coordinate these on a global basis" (Chiesa, 1996, p. 10). This means that various laboratories may specialize on performing specific tasks. From case studies, it can be concluded that some effectiveness problems might arise when experimental R&D is performed by specialized contributors with overlapping tasks. This could mean that the same activity is performed by more than one unit. This was mentioned earlier with reference to world product mandates or world technology mandates. Also, in exploitation R&D, the so-called supported specialization - that is specialization supported by application laboratories or scanning units in other locations - fits this case. These laboratories fulfill a local task, but they cooperate to avoid duplication. Managing laboratories of the case (5) type may be a most difficult task, because it requires very complex organizational structures, as will be shown later, and intense mutual communication.

Case (6) might best be described as "exploitation R&D" with supported units in integrated local laboratories. Thereby, local innovations are combined with global exploitation which can be achieved by involving headquarters in the coordination processes.

This is a case where one laboratory in an organization has been chosen as a competence center for some technology development or that it has developed this capacity over time. It is then given the sole responsibility in this particular field of work, while overriding strategic decisions continue to be made by the headquarters. Thus, decision making is shared between the headquarters and the locals. This arrangement has pros and cons as well. On the one hand, the costs of coordination tend to grow with the number of units to be coordinated. On the other hand, it is possible to develop synergism within the organization. This model is a very prominent one, chosen by 55% of the organizations (Schmaul, 1995, p. 80). As in the hub model, the competence center model is observed in organizations with a small mean number of foreign laboratories (6.9) (Schmaul, 1995, p. 80).

Like other cases, case (8) could be characterized by two types of laboratories. Firstly, experimental R&D performed by specialized contributors that are sequentially related in an overall development task fits this case. Secondly, so-called integration-based laboratories need to be mentioned here. Chiesa states: "Each unit in the network is allowed to undertake its own research initiatives. Within the network, units continuously communicate, interact with each other, and exchange results" (Chiesa, 1996, p. 17). The network might be managed by a network supervisor. The allocation of this type of laboratory to the respective case assumes then that the supervisor is located at a foreign R&D laboratory.

In a network model, laboratories have adopted a particular mode of division of labor among themselves. They enjoy an equivalent status, are given leadership in certain areas, and enjoy a considerable degree of autonomy in decision making in these areas. This may result in a loss of control on the part of the headquarters. The laboratories may compete against each other, and they may strive for more autonomy than may be optimal from the point of view of the total organization. Synergism may not arise easily in this approach. Such "network models" are observed in 10% of all organizations which operate a large number of foreign laboratories (16.5 on average) (Schmaul, 1995, p. 80).

Unfortunately, Chiesa uses data from only 12 companies to develop his classification. Thus, from these observations, little can be said about the relative frequency of occurrence of all cases. Schmaul's results cover only three of the relevant five cases, and, therefore, cannot be used to draw a full picture. Furthermore, the relative frequency of certain organizational arrangements does not signal success. Among other variables, success might also depend on the fit between the structural organization of the laboratory and its immediate company environment. Finally, the relative strength or power of the involvement of foreign units in the coordination or decision processes is not considered in the aforementioned taxonomy.

In conclusion, we find that a differentiation between the locus of decision making on R&D and the locus of R&D performance leads to a number of alternative arrangements for international R&D. These arrangements can be observed in reality. They seem to offer particular benefits and costs that need to be optimized under the prevailing conditions of a particular firm.

5.3.2 Centralization and autonomy

In the last paragraph, we identified two important criteria that help to classify foreign R&D units. With respect to one of these, centralization or decentralization, further details are necessary. Along the continuous scale that measures this variable, we should also be interested in the degree to which R&D laboratories are autonomous in deciding on issues of strategy and operations. In our view, the locus of decision making and the degree to which laboratories can make autonomous decisions represent two variables. Earlier research has treated these variables as only one, or alternatively as highly correlated.

In the most prominent of these earlier studies, centralization and autonomy in the relationship between parent and subsidiary are considered jointly (Behrman/Fischer, 1990, p. 40). These authors introduce five categories of "managerial styles":

(1) absolute centralization, where commitment is imposed by the parent on the subsidiary;

(2) participative centralization, where commitment is reached as a result of negotiation between parent and subsidiary;

(3) cooperation, where commitment is reached by agreement between parent and subsidiary approaching equals;

(4) supervised freedom, where commitment is established by the subsidiary's decision with consulting input from the parent; this was the most frequently used category; and

(5) total freedom.

The perceived degree of parent involvement serves as the criteria for this rather highly aggregated level of categorization. The degree is not differentiated by types of decisions. The authors conclude that the choice of categories is partially determined by the market orientation of the firms. This leaves room for more determining variables.

When coordination activities were discussed above, it was not clear which institution besides the laboratory management was "in charge" or had more power. If headquarters is in charge, the R&D facilities are "headquarters dependent". Behrman and Fischer (1980) call this "absolute centralization". Other authors have interpreted this as a top-down planning approach (Arimura, 1997). If local facilities, for instance production or sales units in a foreign country, are in charge of decision making, we observe "locally dependent" facilities or "supervised freedom" (Behrman/Fischer, 1980). The mixed mode, which is called "facilities with shared R&D decision making" (Behrman/Fischer, 1980) or the mixed approach (Arimura, 1997), is possible as well. It is also conceivable that an R&D laboratory is fully autonomous in determining its strategy and its operations. While this may appear to be an extreme case, such laboratories are called "autonomous facilities" or laboratories given "total freedom" (Behrman/Fischer, 1980). This has been associated with a bottom-up planning approach (Arimura, 1997).

Behrman and Fischer (1980) suggest that U.S. companies prefer top-down approaches, while European companies favor bottom-up approaches. The U.S. management would, thus, favor participative centralization or absolute centralization, while European management favors the supervised freedom or even the total freedom approaches. Even if this observation is correct, it may not hold over time. Granstrand and Fernlund (1978) and Håkanson and Zander (1988) seem to indicate the move towards a mixed approach in European companies. Japanese companies in the electronics industry appear to be moving towards a mixed system as well, either by establishing extra coordination boards or by partitioning projects into two classes. One of these is the class of strategic projects which are planned by and supervised from the headquarters. The other class is that of regional or local projects for which foreign R&D units are in charge (Arimura, 1997). The choices seem to be motivated by learning the problems of international coordination as well as by the increasing number and the rising level of foreign R&D activities. These observations indicate that decisions might be split with respect to the character of projects. If, however, laboratories are not fully designated to either one class of projects or the other, some authority has to decide how much of their capacity to allocate to each class. To the degree that the laboratories decide on this, they are granted a certain degree of autonomy.

Autonomy, then, can be a variable that is distinguished from centrality. Thus, it should be studied separately.

Using interviews in U.S. laboratories affiliated with foreign companies, it is concluded that these "possess considerable autonomy in proposing projects, setting technical agendas, and hiring new staff with these functions being the primary responsibilitiy of in-house scientific and technical staff" (Florida, 1997, p. 99). This indicates that "autonomy" is a multi-facetted construct that should be operational-ized by a number of items.

Operationalization was achieved by defining twelve types of deci-sions that involve some degree of coordination at the laboratory level. These range from making decisions about the R&D program, initiating cooperations, holding strategic meetings about project se-lection and project planning, holding operative meetings, deciding about project budgets and project termination to the assignment of specific personnel.

Four potential decision making units are considered:
– headquarters R&D
– other headquarters units
– local R&D
– other local units.

Data on the decision making units, with respect to the twelve types of decisions, were collected and were used to construct two indices, namely the degree of autonomy (d_a) and the degree of centralization (d_c). The degree of autonomy is defined as the number of types of decisions made in the foreign R&D facility itself divided by the total number of possible decision types. The degree of centralization is defined as the number of decision types made at the headquarters (R&D or some other headquartered unit) divided by the total number of possible decision types.

Data supplied by 31 German companies on 81 foreign R&D labo-ratories were used to measure the degree of autonomy and the de-gree of centralization (Brockhoff/Schmaul, 1996; Schmaul, 1995). Cluster analysis of the responses reveales four different types of labo-ratories. Table 4 summarizes the results.

The data indicate that the four types of laboratories are clearly characterized by the particular value levels of the two indices of cen-tralization and autonomy. Autonomous laboratories (35%) enjoy high autonomy and face a low degree of decision centralization. These are foreign laboratories which by and large define their own activities. Laboratories with shared decision making (30%) only make opera-tive, project-related decisions by themselves while strategic decisions are mostly made at the headquarters R&D unit. 23% of all foreign laboratories are almost totally dependent on central decisions which are either taken at headquarters R&D or another headquarters unit.

Table 4: Laboratory types, centralization and autonomy

Laboratory type	Degree of centralization		Degree of autonomy	
	d_c	Diff.*	d_a	Diff.*
Autonomous facilities (A)	0.05	S,H,L	0.93	S,H,L
Shared R&D decision making (S)	0.35	A,H,L	0.53	A,H
Locally dependent facilities (L)	0.18	A,S,H	0.38	A
Headquarters dependent facilities (H)	0.73	A,S,L	0.36	A,S

* Diff.= indicates those laboratory types from which the particular laboratory differs significantly with respect to either centralization or autonomy.

An interestingly small group of 9% of laboratories has little autonomy, but decisions are made at another local unit, such as local general management, manufacturing or marketing/sales units. These are called locally dependent facilities.

Earlier, three models of organizational structure for international R&D laboratories were identified: the hub model, the competence center model, and the network model. It is of interest to find out whether these models relate in a specific way to the aforementioned four clusters.

Of 26 "autonomous facilities", 23 are found in companies that have adopted a network model. This is in line with the above reasoning. Of 22 "shared R&D decision making facilities", we find 16 that use a competence model in their organizations. This result was expected as well. The 17 "headquarters dependent facilities" as well as the 9 "locally dependent facilities" occur in all three models with no particular center of gravity in any one of them. Looking at the data from another perspective, it becomes evident that these reveal two types of "hub" models:

– one, where the locus of decision making rests with the headquarters, and

– another one, where decisions are made at a foreign unit.

This solution was not evident on first observation.

This finding supports our view that the locus of decision and the locus of performing R&D represent two important variables for structuring international R&D laboratories.

5.3.3 Strategic contingencies and power

In describing the coordination among the R&D units, we relied heavily on centrality and autonomy. Unfortunately, both variables do not fully explain whether a particular foreign R&D laboratory that enjoys a non-extreme level of autonomy can influence the decisions of other units. The question is whether it has power to do so or not. Interestingly, a "network power paradox" has been observed. This means that one organizational unit can have more power than another, although it has less autonomy (Medcof, 1997b). Thus, influence or power is a further dimension that needs to be considered.

In an organizational perspective and in line with the preceding thoughts, power is defined as "the capacity of a subunit to influence the behavior of another subunit" (Harpaz/ Meshoulam, 1997, p. 108). This capacity can have different bases or critical contingencies which are correlated with important characteristics of the work done in R&D laboratories (Medcof, 1997):

(1) *Criticality*, which measures the degree to which the overall success of the organization depends on the work of the individual laboratory. In R&D, the level of strategic significance of the work done in a laboratory is assumed to be positively correlated with criticality.

(2) *Substitutability*, which identifies the degree of monopoly that an individual laboratory enjoys, that is, whether its work could be substituted easily or not. In the competence model there is little substitutability, and if power, that could be derived from this, needs to be checked and balanced, it may be necessary to accept parallel work of the same sort in different places. This could erode the concept of the center of competence.

(3) *Interaction*, which measures the degree to which a laboratory interacts with others. Foreign R&D units that interact with many others are assumed to have more power, ceteris paribus, than those with few interactions. The type of interdependence -

pooled, sequential or reciprocal (Thomson, 1967) - and the direction in the case of sequential interdependence - towards headquarters or towards other foreign R&D units - may be of additional importance in defining interaction as a base of power.

(4) *Immediacy*, which is the time lag between the hypothetical or real stoppage of work in a laboratory and the cessation of the work of the total organization. It is difficult to measure this precisely. In fact, if only minor adaptations to local markets are assigned to a laboratory, it cannot stop the work of the overall organization by withholding its own work. Basic and applied research, when stopped, can have a substantial effect - although not necessary a fatal one - on the total organization after some ten to 15 years. This is demonstrated in company R&D histories (Graham/Pruitt, 1990; Hounshell/Smith, 1989). The long time lag might explain why research managers complain of having only little power in their organizations.

To offer further insight, other power bases may be considered. The network literature offers indicators, such as centrality or periphery in a communication network (Berge, 1966, p. 119), which refine the interaction characteristics. This network measure might be derived from the distribution of competencies among laboratories as well as their access to powerful units (such as top management). The sociological literature considers legal power, expertise, ability to distribute sanctions, hierarchical position, or ecological influence (French/Raven, 1959). In our context, ecological influence is unimportant as it relates to operational aspects only and not to structural and strategic aspects. Expertise is represented by substitutability. The ability to distribute sanctions is related to criticality, interaction, and immediacy. Hierarchical position is deliberately not considered as a variable here, as all foreign laboratories are assumed to take an equal position in this respect.

Thus, the four characteristics mentioned above serve as a good starting point for measuring laboratory power. However, it would be advantageous to aggregate the four characteristics into one construct or index that allows us to measure the power P_i of the i-th laboratory. Assuming M mutually independent power bases (here we have M=4) this index could be defined as:

$$P_i = \sum_{m=1}^{M} w_m p_{m,i}$$

with $m = 1, ..., M$, a weight w_m of each of these characteristics or power bases and $p_{m,i}$ the level of the m-th power base enjoyed by the i-th laboratory. While it can be imagined that $p_{m,i}$ could be measured by offering a seven-point scale to subjects who are invited to evaluate the characteristics for each of several laboratories, measuring w_m could be more difficult. For this purpose, conjoint measurement offers a suitable approach. The weight could be deduced from rating or ranking the power of several, perhaps fictitious, laboratories with alternative levels of characteristics or power bases chosen according to an experimental design.

We know of no empirical study to measure P_i by this or a similar approach. Therefore, we hypothesize that power distributions among laboratories will look like those in Figure 10. These assumptions are only based on plausibility. The network model is characterized by an almost equal level of power for each laboratory, while the hub model assumes power differences between the hub and the spokes. In the center of the competence model, different arrangements can be made plausible. If each laboratory is assigned an almost equally important field of competence, power distribution might be equal to that of the network model. If this is not so, as assumed in Figure 10, a distribution that is more similar to the hub model could arise. Contingencies, such as the type and level of uncertainty involved in the laboratory tasks, could be made plausible as moderating factors in choosing a preferred distribution of power.

Coming back to the network-power-paradox, Medcof (1997b) has made plausible that power can be largely independent from the autonomy of individual laboratories. This is an important message for general managers and for laboratory managers as well. Unfortunately, we do not yet have the knowledge about an ideal distribution of power P_i among all laboratories of a firm that contribute to maximizing performance.

Assuming that the power indices of laboratories from many organizations could be averaged, we suggest that the different types identified before (Brockhoff/Schmaul, 1996; Schmaul, 1995) might be ordered as in Figure 11. With respect to the power axes, the

Figure 10: Power distributions among laboratories in different organizational models

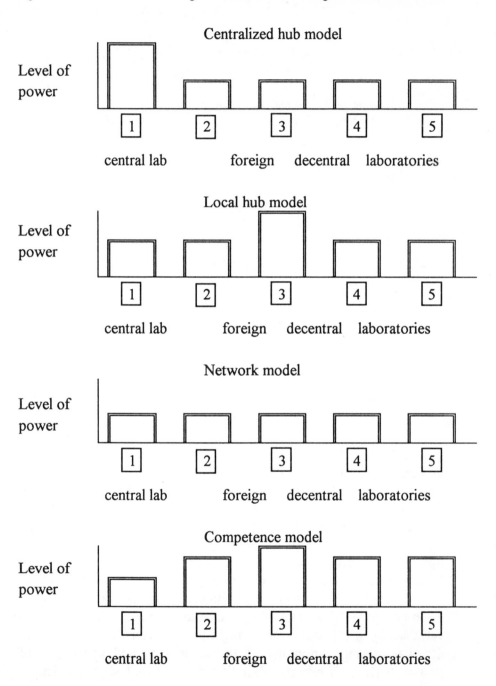

Figure 11: Power and autonomy as laboratory characteristics

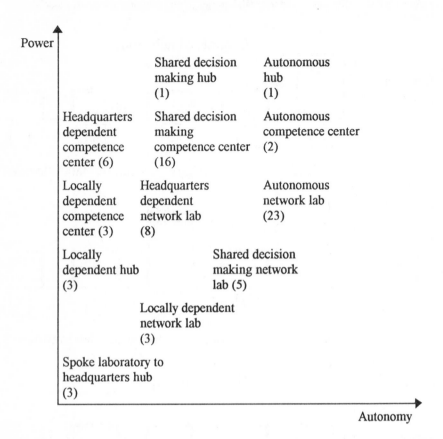

arrangement is based on plausibility derived from descriptive information on the laboratories considered in the study. With respect to autonomy, the index defined in chapter 5.3.2 could be used. The locations presented in the figure are only approximate. The numbers shown in brackets indicate the number of laboratories of a particular type that could be identified (Schmaul, 1995, p. 105). According to this, practice seems to allow for all levels of autonomy, but it favors relatively high levels. Practice also seems to favor medium levels of power. Whether this is correlated with success remains to be seen.

To conclude this part of our presentation, we have identified three variables by which laboratories might be characterized: centrality of decision making, autonomy of the laboratory, and its degree of power. It has also become clear that these are not strictly independent of each other. Only sparse information is available on the relative frequency with which combinations of specific values of these variables are chosen. Whether particular variable levels relate to success remains to be seen. It is even unclear whether management has in the past deliberately chosen particular combinations of variable levels to achieve success. Occasional interviews could not identify high awareness of the possible combinations of the variables. Thus, it did not occur that by influencing the level of the variables, laboratories themselves (via some of the power bases), as well as general management (via some other power bases, centrality of decisions, and choice of an acceptable level of autonomy) influenced the status and possible success of laboratories.

6. Coordinating internationally dispersed R&D laboratories

6.1 Interface management

In a questionnaire study among the world's largest companies, Pearce and Singh find that only a quarter of all laboratories is closely coordinated with the parent company, while 22% of the companies do not see any coordination (Pearce/Singh, 1991, pp. 198 et seq.). Most of the coordination is approached systematically (68%), some in an ad hoc fashion (25%), and the rest infrequently (7%) (Pearce/ Singh, 1991, pp. 198 et seq.). This leads us to think that quite a number of firms face severe problems in managing the interfaces among laboratories, and between laboratories and the other departments of a company.

Organizational interfaces are observed when:
- two or more individuals or groups need to cooperate to solve a particular problem or task;
- these problems or tasks are neither coordinated by market pricing mechanisms nor by hierarchical advice;
- these individuals or groups cannot refer to a superior person as an arbitrator in their operational problems and conflicts.

The necessity of interaction across the interface might breed conflict. The types of conflicts depend, in part, on the kind of interaction. Sequential interaction is liable to be affected by differences in perceptions and knowledge resulting from informational asymmetries or cultural differences that impact the frame of reference of individuals. Thus, conflicts arise from lack of information or lack of consensus. Pooled interaction occurs where resources have to be shared and some priorization scheme needs to be in place to settle resource conflicts. Reciprocal interaction might lead to both types of conflicts. Different perceptions and perspectives about objectives as well as about status and roles of participants in R&D networks that spread across national boundaries can be a source of substantial irritations. These psychological and sociological aspects of organizational interfaces make their management more difficult than dealing with technical interface problems.

One idea might be to eliminate all organizational interfaces. Except for the one person organization, there is no way of avoiding interfaces altogether. Another alternative, utopian socialism, whereby every member of a society is trained and encouraged to take any job, has not proven successful, not even at relatively unsophisticated levels of societal development. Thus, in sophisticated R&D organizations that need to interact among each other and with their external and internal corporate environments, the threat of conflicts is ever present. Eventual conflicts can reduce the efficiency and effectiveness of the organization. This is why interfaces need to be managed (Brockhoff, 1996). Unfortunately, empirical research has identified substantial levels of conflicts in international R&D organizations.

Interviews in seven British firms with international R&D activities have indicated substantial ineffectiveness, which is primarily explained by delayed market entry and the choice of improper objectives in R&D work (von Boehmer, 1995, p. 121). The latter, in particular, could be attributed to the insufficient market knowledge of developers, a substantial number of shelved ideas, and some bootlegging (von Boehmer, 1995, p. 122). The same ranking order was found for these reasons in a study of communication problems in German laboratories. However, the ranking order was topped by the specific German problem of a drive towards overperfection (Brockhoff, 1990, p. 34).

A large number of shelved ideas and bootlegging activitites is an indication of a split between the decision makers in management and the R&D staff in terms of their R&D objectives. Which of these groups of ideas serves the companies better cannot be determined. But, the existence of the differences is indicative of weaknesses. The British managers complained that information exchange is guided more by mutual sympathy than by the necessity for this exchange, and that, in particular, the feedback from marketing was too little (von Boehmer, 1995, p. 123). Interestingly, these complaints took top priority, as well, among the German managers in the strictly national management of R&D (Brockhoff, 1990, p. 37). These findings indeed indicate substantial interface problems. To overcome these problems, management needs to employ particular instruments for coordination.

The coordination of internationally dispersed R&D activities draws on a large number of such instruments. In hub models, coordination might be achieved by using the hierarchy that exists between the hub and the spokes. In other models, such as the competence or the network models, there is no clear hierarchy between the different laboratories and the specifics of new knowledge, as their outputs make it difficult to adopt a pure pricing mechanism to coordinate activities. Thus, we observe a definite need for coordination by managing interfaces in a non-hierarchical and non-price regulated environment. For this purpose, hybrid instruments are available in large numbers.

The problem to be solved is that of a special kind of interface management. It has been described as the need to achieve "global localisation" (Sommer, 1990). This means making use of local strengths and making managers aware of global possibilities and needs. At Sony Corp., managers try to achieve this by using a specific way of international education for regional managers without interfering with the cohesive strength of the company. At Schering AG, where R&D is performed at headquarters and internationally dispersed laboratories and where decisions are made by involving headquarters as well as foreign units, there exists a very complex structure of overlapping groups (See Figure 12). This structure evolved from the earlier experiences and changing needs of an increasingly more internationalized as well as cooperative mode of R&D. It is backed up by institutionalized planning processes and formal as well as informal communication.

There are two groups of instruments for interface management that could be chosen if interfaces cannot be avoided. One group is complementary to the activities performed by R&D people on either side of an organizational interface. A second group of activities is only performed by specific interface people, mostly as their major professional activity or as a full-time job. These alternatives are shown in Figure 13.

Let us first consider interface management as a full-time job. It could be performed by special assignment or as a result of having learned it by a process of socialization. Examples for this are men-

Figure 12: R&D planning groups at Schering AG

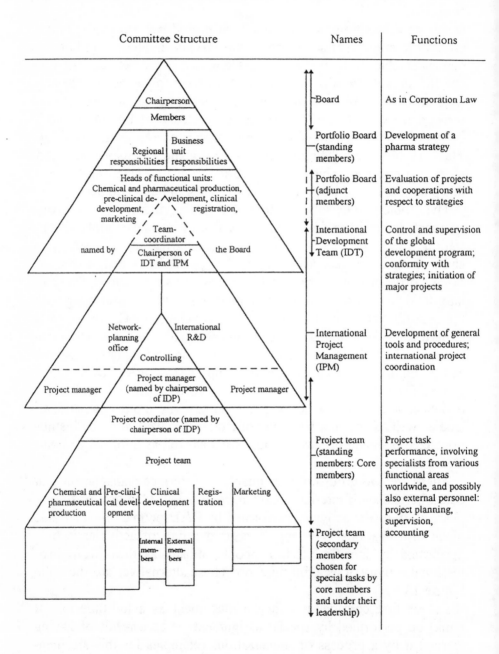

Committee Structure	Names	Functions

Source: Drawn on the basis of company releases. Not authorized by Schering AG.

tioned in Figure 13. At the project level, project leaders, champions,or process promotors might be appointed to coordinate contributing units which could be various laboratories in different countries. This is particularly relevant if network structures or competence center structures represent the R&D organization. On a more permanent basis and at a higher level, coordinators or liaison people might be appointed. For example, Siemens AG appoints "key account managers" who are laboratory employees. They are charged with the task of identifying, for particular internal customers, the sources of technologies that could be integrated in products that these customers need. These sources might be found in one or in several internal laboratories.

Figure 13: An overview of interface management

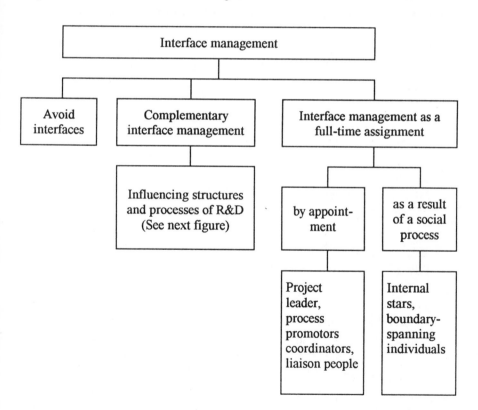

In addition, complementary instruments of interface management might be applied at any job. These encompass a large number of instruments. We present some of the most frequently mentioned instruments in a systematic fashion (Figure 14). (For an alternative presentation see: Reger, 1997, with examples from Philips and Hitachi).

On the one hand, complementary interface management can make use of instruments which might either supplement the structural characteristics of the organization or the processes that are followed. On the other hand, the instruments might either concern individuals immediately or refer to non-personalized activities. Using these criteria, a two-by-two table can be developed into which the different instruments can be placed.

A major question is: under which circumstances do particular complementary instruments promise best results? Unfortunately, an algorithm does not exist by which the most successful instrument or combination of instruments could be selected. Earlier research has shown that four groups of interface characteristics should be observed and identified in selecting particular instruments or in rejecting other ones.

(1)   The level at which the interface problem arises should be considered. Whether this is at the laboratory level or at the level of an individual project is of some importance for eliminating some instruments or choosing some others. Case studies of Philips and Hitachi demonstrate very nicely that different instruments are chosen at different levels of coordination (Reger, 1997, pp. 100, 114). For instance, at a high level, the coordination of a laboratory mission with that of other laboratories and the overall strategy of a company require instruments, such as the development of a common vision and joint committees, to define and to oversee the division of labor. Installing a common company culture to facilitate cooperation is another alternative. However, it needs to be applied with great care. Research has shown that in U.S. laboratories that report to Japanese firms, a "U.S. culture" is preferred over a "Japanese culture" (Florida, 1997). The application of a Japanese headquarters culture to achieve coordination is reportedly rejected by laboratories outside Japan. Such attempts create communication barriers and distress to those

Figure 14: Instruments of interface management

Orientation	Personal	Impersonal

Structure	Teambuilding in
	
	  – committees,
	  – new product groups,
	  – project teams,
	  – task forces;
	
	staff work;
	
	matrix organization;
	
	readiness to integrate various subcultures;
	
	consideration of cooperation supporting characteristics in choosing personnel;
	
	inventory of capabilities.
		Distance reduction by
		
		  – decentralization,
		  – spatial arrangements;
		
		planning with decompositioning algorithms;
		
		programs and program documentation;
		
		definition of inter-unit joint product platforms;
		
		transfer pricing;
		
		data banks on projects and facilities.

Process	Joint formation of objectives, goals or norms; development of shared visions and strategies for the organization; avoiding extreme solutions for partial or functional objectives; use of incentive systems honoring international cooperation and coordination; learning about differences in activities, for instance by participation in job rotation or further education; internal information meetings and support of personal contacts.	Securing up-to-date information on plans by information exchange; effective reporting rules and standards, for example, road maps of technological developments; networking with milestones; simultaneous engineering; job rotation programs; design of programs for developing a company culture supportive to cooperation (including incentive programs).

who are asked to adopt them (Reger, 1997, p. 132). At a much lower level, we consider software development between shifts in laboratories that are strategically placed around the globe such as to make possible continuous daytime work on a joint task. The coordination between the shifts is supported by a common understanding of the task, regular information exchange, and possibly some job rotation. It might be necessary to perform a number of projects of this type to learn how to handle them. To reduce the risk of doing harm to the business of a company if such learning appears at a too slow rate, experimental internal projects might have to be defined and worked on before engaging in a market-related project. The success of such learning experiments is dependent on a low rate of fluctuation of the personnel involved.

(2) As indicated above, the type of interaction is of interest because it may explain particular types of conflict. If the laboratories are in conflict over the joint use of a scarce resource, such as finance, access to a large instrument, etc., this conflict might be solved by adopting a pricing scheme for this resource. If one laboratory delivers certain results or resources to another laboratory, conflicts that arise from asymmetric information and differences in perceptions need to be addressed. This cannot be done by pricing. However, information exchange by way of meetings, joint committees, etc. might be helpful. If there is a reciprocal exchange of resources or new knowledge between laboratories, the development of joint visions might help to reduce the conflicts that might arise otherwise. A number of techniques are available which help to develop joint visions, goals or norms from diverging views held by members of different groups. Elaborate experiences have been collected from applications of variance-reducing, anonymity-prevailing, structured and repetitive communication processes. The Delphi-Method probably is the most well-known procedure of this kind that has found many and diverse applications (Linstone/Turoff, 1975). A non-anonymous and non-repetitive procedure is known as the Analytic Hierarchy Process which has found an application in coordinating project partners (van Rossum, 1998). Both approaches are rightfully criticized for certain defects; however, they cannot be criticized for not supporting convergence on conflicting views.

None of the methods can guarantee to achieve full consensus. It is, therefore, necessary to establish "freezing points" when a certain result of discussions cannot be revised, as well as an arbitrator who is in charge of settling an issue or determining the inability to come to terms. Irrespective of the methods that might be applied, we find that the type of interaction codetermines the choices of interface instruments.

(3) The reason for the existence of an interface should be considered. In interfaces that arise from a division of labor because of specialization, one should favor the selection of instruments that are different from those that appear to be effective, if parallel work on the same tasks is chosen to speed up processes or if the size of the task exceeds the capacity to handle it in one place. In the first case, instruments supporting effectiveness are more important, while in the second case efficiency maximization at a predefined quality level should be stressed.

(4) Task characteristics need to be considered because they could be of importance to the organization, the degree of novelty, the possibilities of structuring the task and separating parts of it, and to the ability to codify information needed for task performance and originating from it. The latter criteria have been used to explain the choice between interlocal projects and intralocal projects (Gassmann, 1997a, p. 145, 148 et seq.). However, this choice is interdependent with the selection of more or less effective instruments of interface management.

If an interface-crossing process is of very high importance to the organization, cost-intensive instruments can be justified. These could be full-time coordinators or liaison personnel. If, however, the process is of only minor importance, complementary instruments that are used while performing the respective task should be adopted. Finally, multi-technology projects may call for very particular coordination instruments that are not used in other types of projects (Reger, 1997, p. 122 et seq.).

If the task can be structured and separated easily and if information can be codified, more non-personalized instruments of interface management can be chosen; however, if these characteristics are missing, more personalized instruments need to be applied because observing the people involved in the process allows a transfer of the more implicit or tacit parts of the knowledge base.

The time-domain under which a common task has to be performed is a very important characteristic in international R&D. Contributions that originate in different time-zones need to be coordinated. They may be a source of improved efficiency in well-structured, hierarchically organized tasks. They might be a source of escalating conflicts if they cannot be collected, disseminated, and reacted upon concurrently. Traditionally, this was achieved by face-to-face discussion. In the next chapter, we shall further explore to what degree modern communication media might serve this purpose. We doubt that these media can be a full substitute for face-to-face communication.

Applications of the criteria have shown that it is easier to eliminate certain instruments from further consideration step-by-step than to adopt particular ones. Elimination narrows down the list of potentially successful instruments. Finally, very few instruments will survive the successive steps. It is then necessary to decide whether these should be applied in combination.

A further test of the feasibility of the application of the remaining instruments is whether these are acceptable to the different cultures into which the laboratories on either side of the interface are imbedded. Case studies have shown that the evaluations of certain instruments might be influenced by cultural differences between nations or regions. Thus, contrary to Japanese companies, "Western" multinational firms appear to more easily accept the structure-related impersonal approaches (including internal markets and transfer pricing), while they less frequently use structure-related personal instruments (Reger, 1997, pp. 117 et seq.).

Any selection of instruments that appears to be feasible needs to be closely monitored in order to develop experiences on the consequences of the application. These can be evaluated and might lead to possible revision of an earlier decision on the instrument selection. A flow chart summarizing these procedures is presented in Figure 15.

Figure 15: A schematic approach to the selection of instruments for interface management

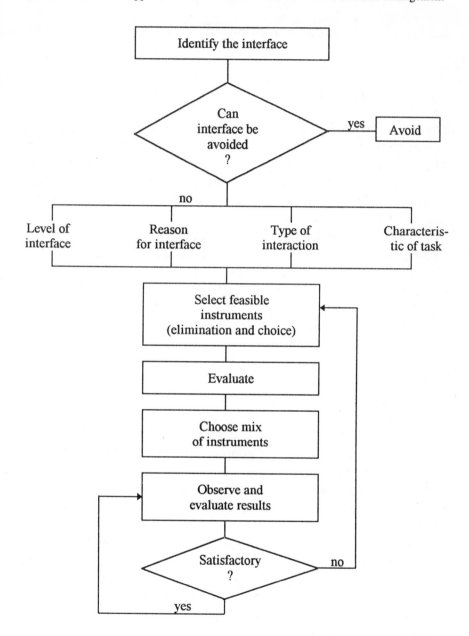

6.2 Communication technologies

6.2.1 Conditions of use

The key to managing internationally dispersed laboratories within one organization is communication. This could be communication about strategic decisions, supervision and control, portfolio management, project performance, skills of personnel, etc. Communication technologies might affect the contents and the frequency of message exchanges among laboratories and individuals.

In a sample of foreign laboratories based in the U.S., quite intense communication frequencies could be identified. Close to 78% of the laboratories report to a sister R&D facilitiy and 63% report to corporate headquarters: ". . . more than 40% of the respondents indicated that they report to a sister R&D facility on a daily basis and 30% do so on a weekly basis. Roughly 35% of respondents indicated that they report to corporate headquarters on a daily basis and 30% do so weekly" (Florida, 1997, p. 98). It is also said that much of this communication is concerned with administrative and coordination functions. Thus, the exchange of project-related problems, solution steps or intermediate results, and the request for advice or help might not have been captured in the study mentioned. These activities need to be considered as well. At the laboratory level, De Meyer (1993) reports on substantial travel budgets and meetings for coordination and control that can take up 2.5 to 5 days per month and researcher. Empirical research indicates dissatisfaction with this state of communication.

Modern communication technologies might support globalized R&D in many respects, ranging from enabling better managerial decisions on R&D to more efficient and effective technical decisions and controls (von Boehmer, 1994; Chakrabarti, 1994). Even the conduct of meetings can be made easier, such that off-line discussions become possible (Weatherall/Nunamaker, 1995). In general, one might say that knowledge transfer is the aim of communication. This transfer might be facilitated by modern communication technologies. This being the case, several questions arise:

- What kind of messages is facilitated by such technologies,
- what is the balance of benefits and costs, and
- is more centralization or more decentralization likely to occur?

If one wants to answer the first question, a systematic look at the occurance of knowledge transfer appears to be necessary. Three dimensions have been suggested to arrive at a taxonomy of cases:

(a) the presence or absence of a perceived relevance of knowledge as it contributes to the solution of a problem (having the ability to see how the means corrspond to the ends);

(b) the willingness to articulate or retain knowledge irrespective of its possible relevance for problem solution; and

(c) the economic efficiency of knowledge articulation (which might be precluded by the excessive costs of documentation, a lack of language for proper expression of knowledge, etc.).

Combining these criteria leads to the eight cases shown in Figure 16. These are discussed without going into too much detail.

Figure 16: Problems of knowledge transfer

| | Perceived relevance of knowledge for problem solution | | | |
| | present | | absent | |
	Knowledge articulated	Knowledge retained	Knowledge articulated	Knowledge retained
Articulation efficient	(1) Ideal case of knowledge transfer with media use	(3) Opportunistic behavior to build a power base by not-transferring	(5) Recipient of knowledge is unaware of its problem-solving potential	(7) Two-sided overlooked knowledge potential (classic case of tacit knowledge)
Articulation non-efficient	(2) Non-economic transfer of knowledge perceived as relevant (recipient might learn at lower cost on his own)	(4) Economical non-transfer of knowledge (tacit knowledge)	(6) Non-economic transfer of knowledge that is not perceived as relevant and therefore not used by the recipient	(8) Unawareness of transfer opportunities without particular disutility of non-transfer to the recipient

Source: Adapted with changes from Rüdiger/Vanini, (1998).

Case (1) is immediately amenable to media-supported knowledge transfer. Cases (2), (4), (6), and (8) should preclude transfer with the media presently discussed, although in Cases (2), (4), and (6) richer or less expensive communication, such as personal contacts, might lead to beneficial transfers. Cases (3) and (4) need additional incentives in order for the owner of knowledge to articulate his or her knowledge and to make it transferable. However, the cost of providing the incentive might be higher than the cost of learning by the potential recipient without any transfer. Case (4) describes one of those situations that have been called "tacit knowledge": knowledge is retained because it is not considered worthwhile to transfer it. Cases (5) and (6) have led to good documentation and also to transfer of knowledge. However, in Case (5) the recipient receives a diamond that he identifies as carbon, while in Case (6) the transfer of the diamond is so expensive that it does not get used for two reasons: it appears useless and too expensive. In Case (7) it might be very beneficial to transfer what appears to both sides to be a piece of carbon, but when cut and polished would become visible as a diamond. This potential remains undiscovered. It is the classical case of tacit knowledge. A joint look at the finding and its discussion might soon lead us to recognize its true (problem-solving) nature. This discussion sets free associations and combinations.

Thus, perceptional as well as cognitive problems have to be solved to make knowledge transfer feasible. Feasibility might depend on the characteristics of communication technologies. From observations of information exchange within multinational firms, it is clear that even advanced electronic communication media cannot fully substitute for intensive person-to-person contacts. Even so, some of the expensive technologies which offer very rich communication possibilities are retained.

To what degree knowledge transfer by modern communication technologies is considered to lead to benefits or costs is very difficult to decide. At first sight, one observes a dramatic decrease of communication costs per message over a given distance. Efficiency of R&D processes could also be affected by allowing for higher speed of developments. Speed, high capacity, and the low cost of information transfer have contributed to the possibility of continuous research and development activities. This is made possible by working shifts on a project in strategically chosen places around the globe and by forwarding the results of one shift worked in a particular laboratory to the next laboratory at day's end, which starts its day from this information. This working pattern has been used in software development, the automobile industry, or chemical research. Among the criteria for the success of such procedures is the ability to split a problem into independent subtasks (as in chemical research), to communicate very clearly the starting point for the necessary next steps as well as their direction, and to integrate either of these criteria within a commonly followed set of objectives.

A further benefit of modern communication technologies is their ability to enable very rapid learning. The "exposure to sources of knowledge in different countries is important. But to be effective, one has to create mechanisms on an international scale to diffuse, validate and integrate the new knowledge across the whole network of laboratories. And diffusion, validation and integration are heavily determined by the quality of the formal and informal communication system", says De Meyer (1993). Thus, laboratories need to develop the potential to identify relevant knowledge, to absorb relevant knowledge, and to disseminate relevant knowledge to other laboratories or customer departments (Brockhoff, 1997). However, presenting benefits does not mean giving the full story.

Research on long-distance cooperation ("telecooperation") has concluded that participants need to develop certain qualifications to make use of potential benefits. People are considered the most important resource to enable organizational learning and innovativeness. They should develop basic competencies in the use of communication media, have communication abilities, be flexible, ready to learn, able to self-manage, have social skills and the ability to work in teams; they should be able to work creatively and autonomously in solving problems and making decisions (Reichwald et al., 1998, p. 217). This is an extremely demanding program that will be met by only a few people. We mention it primarily for the reason that it draws attention to a number of non-technical cost factors.

A second more general cost-benefit factor is defined by the initiative of communication. It is reported that the effectiveness of processing information is related to performance measures (Fischer, 1980). This relationship requires a bit of differentiation. Experimental research has found that performance is not affected by the increasing supply of information. Rather, if increased relevant information demand is met, it gives support to decision quality (Brockhoff, 1986). The more relevant information demand can be, the less communication technology constrains quasi-natural modes of questioning or reporting. Therefore, modern communication technologies hopefully contribute to improving relevant information exchange.

The lower cost of communication per message and per mile could be interpreted to make cooperation feasible over long distances. However, the frequency of communication diminishes as the distance between laboratories is increased, and as the time needed to process customer's product specifications to laboratories and back expands with increasing distance (Hough, 1972; Allen, 1977; Howells, 1990). At first sight, this seems to apply to personal contacts only. However, if the use of electronic media depends on the psychic distance between senders and receivers, and if this psychic distance is correlated with geographical distance, then similar observations might be made.

At any given level of communication frequency, lower communication costs might result in accepting longer distances. This could be particularly annoying if personal contacts continue to be important. As shown in Figure 16 (above), this could indeed be so in order to make the transfer of some tacit knowledge feasible. Even in recent years, laboratory structure, for example in the automobile industry (BMW's research and engineering center) or the biotechnology industry (Elf's laboratory in France), was deliberately chosen to facilitate personal contacts and interactions.

The communication systems used to support the various tasks arising in R&D laboratories produce quite substantial costs of coordination and control. These might more than offset potential benefits from such an activity. To support communication, more and better documentation of project progress is helpful as is an exchange of personnel between laboratories. The first prerequisite might lead to more bureaucracy. Controlling the access to jointly used data banks of projects, capabilities, methods, or data is a largely unsolved problem within firms (Gassmann/Boutellier, 1997, p. 29). The solution has to balance a need for secrecy with the benefits of broad information. The personnel exchange could lead to more travel in spite of the use of electronic communication.

Communication costs are context-specific in an external and in an internal way. In the external way, it should also be noted that the level of communication quality and costs is not standardized internationally. This could mean that certain arrangements of R&D dispersal are economically feasible between a particular group of nations and not between others (Reddy/Sigurdson, 1997, p. 359). This might explain the contradictory observations or different decisions made in previous years when relatively high communication costs were in effect and those made during recent years when lower costs prevail.

In the internal view, media and task characteristics come into play. It is obvious that the characteristics of the media which were used determine the effectiveness of the support for the various tasks. However, communication needs vary over the life of a project. The use of communication media that meet these needs would, therefore, have to change accordingly.

The increasing complexity of technologies and more heterogeneous customer requirements that result from serving more and more remote markets could increasingly require not just more, but also richer communication. This would then drive up communication costs which could consequently more than offset the cost reduction offered per information exchange by modern information technology.

Where benefits and costs can be observed, companies look out for their optimal balancing. As mentioned in the first chapter, more internationalization is observed. This might indicate that, in general, benefits outweigh costs. The development of networks of R&D co-operations across national boarders has greatly expanded.

Howell suggests that: "The evolution and development of such networks will not only have the potential to improve the efficiency of research generation and the interchange of information within and between R&D and other corporate functions, but will alter the way R&D activity is organized and located" (Howells, 1990, p. 143). He hypothesizes that the dramatic decrease of communication costs and the increased "richness" of communication via electronic media in recent years has contributed to further dispersal of R&D units. The enhanced richness, in part, results from the fact that many of the new systems have evolved from the communication needs of researchers. This need has led to a number of internal communication networks that exist beside the publically accessible computer networks (Howells, 1990). They have been used to support the further specialization of laboratories and to reduce the unwanted duplication of projects. Besides influences on structural decisions, there is also speculation that communication devices might support creativity.

There are a few examples where experience has taught firms that the hypothesized promises of beneficial internationalization could not be met. In the late 1970s, Glaxo used its internal communication network "to integrate the research function . . . under a single company-wide unit" (Howells, 1990, p. 142). After time-consuming and costly attempts at designing the Contour or Mondeo model by teams located in Cologne and in Michigan and working on the same project, Ford is quoted as having re-centralized its design teams for new passenger cars (Patel, 1996, p. 46). It is reported that groupware (software to support group processes like brainstorming) is hardly used, mainly because such group processes are too indiscriminating and produce too much "noise" (Gassmann/Boutellier, 1997, p. 29). This may not be generalized to all types of R&D work or to all organizational structures. It is, therefore, important to look for more experiences gained by internationalized R&D projects. As of today, none of the communication media achieves the richness of personal exchange. The more important this is, for instance, in transferring implicit knowledge, culture-bound body language, etc., the more personal contacts are needed. As trust built by personal contact does not seem to last, renewed contacts appear to be necessary. This might explain why travel occurs in addition to electronic communication.

The difficulties in measuring information system benefits in economic terms and the cost of adapting the systems to the rapid technical change in this area are concerns of R&D managers (Chakrabarti, 1994). Thus, at present, there is no easy, standardized answer to the question whether modern communication technologies favor or disfavor all types of R&D work.

6.2.2 Structural interdependencies

Organizational structure, cultural influences, and laboratory mission can all interact with communication activities.

Unfortunately, it is not known how the division of labor among laboratories, their organizational structure (hub, center of competence, or network), the degree of centralitiy and the degree of autonomy of foreign R&D units impinge on the quantity and quality of communication.

One difficulty in measuring influences on communication results from possible interaction with yet another set of variables representing national cultures. Cultures appear to shape preferences for communication channels.

In one study, it is found that U.S. laboratories in the biotechnology industry that report to Japanese headquarters have a quite different information behavior as compared with those reporting to European headquarters (Stock/Greis/Dibner, 1996). In the subsidiaries of the European firms, significantly more personalized communication was observed among scientists of different laboratories within one company as well as between scientists of one laboratory and managers of another laboratory. Japanese parent companies showed a pattern where scientists communicated with their immediate superiors in management, and they, in turn, exchanged information with the management of other laboratories. As mentioned above, variables beyond culture might contribute to an explanation of these patterns. For instance, the laboratories of European parent companies are more frequently charged with applied research and development tasks while the laboratories of Japanese parent companies more frequently perform basic research. Therefore, "cultural, organizational, and technology-related characteristics of the parent-subsidiary" relationship may all come into play to explain these observations. However, some critical remarks need to be added. Looking at merely nine laboratories generates too few observations to measure the effects of all of these variables. Furthermore, no performance measures are available to evaluate the different patterns. Similar caveats are formulated by observing international projects within companies in the pharmaceutical industry (Carbonare/Völker, 1996, p. 61; for practical experiences on managing international projects see: McNichols, 1996).

We expect that communication needs vary with the mission of a laboratory. Data on 186 foreign laboratories of companies based in Germany, Great Britain, and the U.S. support this (von Boehmer, 1995, p. 75). The laboratories could be grouped into five clusters. Communication activities are described by the number of links with internal or external groups as well as by the intensity of cooperation with other R&D units within the firm. Two observations emerge: the more narrow the scope of the R&D tasks, the smaller the number of links with other groups, and the more research-oriented the mission, the more intense the cooperation with other internal R&D units (See Table 5).

Table 5: Laboratory mission and communication activities
(Deviations of cluster means from overall mean value)

Mission	R(%)	# links	Coop.
All-round development with regional focus	47	0.23	0.20
All-round R&D with global focus	9	0.23	0.16
Applied research	20	0.04	-.20
Cooperative basic research	5	-.34	0.78
Application engineering with regional focus	19	-.63	-.56

R(%) :   relative share of the number of laboratories in the cluster;

links :   deviation of mean number of communication links to other groups in the cluster from the overall mean;

Coop. :   deviation of the mean intensity of cooperation with other R&D units within the firm per cluster from the overall mean.

Source: von Boehmer, 1995, p.75.

Broad R&D missions combined with some research missions are correlated with an above average number of communication links and intensity of internal communication. Some of this is needed to coordinate R&D programs and projects. How such coordination is achieved and by whom is not a trivial question. Agency problems of hidden information type and hidden action type might be observed. In fact, companies complain about problems of meeting time goals because laboratories contributing to subtasks of international projects attribute lower priorities to joint projects and higher priorities to their "own" projects. Such problems arise if multiple lines of autority or advice are in conflict or if perceptions on urgency of tasks are not identical in different parts of an organization. Given the very particular character of new knowledge and the particularities of the processes of its generation, some central authoritiy is likely to assume responsibilitiy for coordination.

While such an authority is not necessarily located at a headquarters, it needs to be empowered to make decisions. The authority might be vested in project managers at the project level or in steering committees at the program level. Expertise or legal regulations are the most likely power bases. However, particularly when the network model has been installed (See Figure 4, above), it is necessary to appoint some coordinating institution. This is what we find in many of the largest companies with internationalized R&D, even though they may not have adopted a strict network structure for their laboratories.

Reddy and Sigurdson (1997, p. 358) observe that corporate headquarters must develop capabilities to monitor internal and external technological developments as well as expertise that could deal with it. However, the authors do not refer to particular experiences. Sony Corp. has introduced a system whereby the responsibility of formulating and implementing the global R&D strategy rests with the Chief Technology Officer of the headquarters in Japan while chief technology officers in Europe and America take responsibility for formulating regional R&D strategies (Arimura, 1997). This "zone CTO management system" has created an extra level of hierarchy. Kuemmerle (1997, p. 62) reports on "technology steering committees" that have "five to eight members and include managers with outstanding managerial and scientific records and a range of educational backgrounds and managerial responsibilities . . . In many cases, members include the heads of major existing R&D sites." From the further description, it is obvious that the primary role played jointly by the committee is that of a "power promotor" as described in the project management literature. Its members can mobilize resources at short notice, and they can certainly also withdraw resources. They participate in managing and supervising the R&D programs of their company, and they take responsibility in developing the necessary steps from which a cohesive R&D culture can grow within their company.

Centralized control of network-type international R&D laboratories can be achieved by establishing a virtual or real "central technology unit" (CTU). It could operate similarly to a strategic business unit (SBU) with profit responsibility by issuing R&D contracts to the different laboratories, absorbing their results and licencing the results to the different business units or even external customers. Nestec is a real CTU for Nestlé SA (de Meyer, 1988). Unless SBU representatives can influence decisions taken by the CTU management, there is a real danger that its remoteness from the market place will become a major restraint on company growth and performance. Bringing a CTU into life might involve some financial and tax problems (Boutellier/Kloth/Bodmer, 1996) that need attention. A major benefit, however, should be expected from the ability to coordinate efforts among laboratories. Observations about an additional level of coordination were also made by Gassmann (1997b, p. 160 et seq.). His research indicates that international R&D requires coordinating units such as a steering committee, a systems coordinator or a core team representing various sub-projects. The particular choice of these units is related to project characteristics. These project characteristics are:
(1) incremental versus radical innovation,
(2) autonomous versus systemic type of task,
(3) use of redundant versus complementary resources,
(4) use of explicit versus implicit knowledge.

Interlocal projects, which involve R&D units located at different sites, are chosen if innovation is incremental, tasks are relatively autonomous, resources are redundant, and mostly explicit knowledge is used. To what was said earlier, we add the ability to split a task into separate sub-tasks. Intralocal projects, which can be performed at one site alone, are chosen if innovation is radical, projects are systemic, resources appear to be complementary, and mostly implicit knowledge plays a crucial role in developing the project. The combination of the project characteristics led to 16 different cases. Four cases are mentioned explicitly while the remaining 12 cases are not related to particular project coordination mechanisms. From a discussion of case material, it is made plausible that intraorganizational projects could be handled well by centralized venture teams with a steering committee while interorganizational projects could be managed by decentralized self-coordination, with a central unit setting priorities only implicitly by assigning budgets to sub-tasks.

Certainly, the terms inter- and intraorganizational do not relate directly to an international dimension of project teams. This dimension is addressed if national projects and transnational projects are differentiated, that is when the former involve only people from one national laboratory while the latter involve people from two or more laboratories located in different countries (Gassmann, 1997b, p. 33). Using both criteria, we arrive at a 2x2-table (Figure 17).

Figure 17: Project characteristics and coordination committees
   (according to Gassmann, 1997b)

	Intralocal projects	Interlocal projects
National projects	Systems representative as interface manager	Core member team composed of project representatives
Transnational projects	Central venture team with steering committee	Decentralized self-coordination

So-called globally oriented innovators (Gassmann, 1997b, p. 42) are supposed to favor interlocal, transnational projects and should thus be expected to choose decentralized self-coordination over other forms of coordination. This would correspond with a network organization. This solution is not easily related to the observations about a tendency towards re-centralization among multinational companies which have been the prime respondents to both Schmaul (1996) and Gassmann (1997b). Either the hypotheses developed from these cases prove that they do not hold for larger samples of companies, or that these companies do favor product development for local markets (which is not radical, uses redundant resources and explicit knowledge, and can be handled autonomously), or that more characteristics have to be taken into consideration in explaining the results.

A problem unresolved by these approaches relates to the coordination of other business functions such as production or marketing. The global distribution of competencies in developing particular new technologies may diverge geographically from the competencies in production or those in marketing efforts that are needed to promote successful new products. Thus, different networks of competencies need to be knitted together.

6.2.3 Constraining communication

Modern communication media invite communication and make it relatively easy. Companies have spent substantial amounts of money on building information repositories. By their very existence, these repositories might trigger communication because they provide anchors on which one might want to comment or views one might want to influence. This could be particularly so if commentators are relieved of the binding forces of joint work in natural groups of people. Therefore, it might not be wise to make information available to as many people as possible.

A first constraint to be established is that of defining user groups and the type of information these groups might access. Most importantly, "internal" and "external" groups need to be differentiated. This is getting more difficult, as more internationally dispersed laboratories are becoming involved in cooperations with other institutions. Password systems need to be developed to protect proprietary information.

A second constraint to be established is one of freezing points for changes or termination-points of debates. Rules of order, like Robert's Rule in face-to-face discussion, need to be developed and applied. Application may be achieved automatically, in some cases, or by a discussion leader, arbitrator, or "benevolent dictator" in other cases. These constraints put time-constraints on purposeful communication.

A third constraint is that of accepting only "relevant" communication. Unfortunately, in problems of uncertain outcomes and uncertain solution algorithms, there do not appear to be testing devices that could be applied to detect irrelevant communication (except, perhaps, repetitive communication). In addition, the value of information is perceived differently by different addressees. Assuming that results of communication among a group of laboratories are shared, one might think of pricing the communication differentially. In the absence of normative models or empirical information, it can only be speculated that it might be advantageous to charge lower prices for late, non-domain-specific communication.

The three constraints discussed indicate that using a communication media which is free for everybody and which is available at any time, disregards not only potential information overflow but also issues of competition.

7. Measures of success

7.1 Multiple perspectives

Management of foreign R&D units proves to be a difficult task. This has to do with the observation that, to date, little "hard" data on this topic are available that could guide management in its decisions. For this purpose, evaluations of R&D success would be most important. Only very few studies seek to measure success. Technical success can be measured at the project level as well as the program level. However, this is only a necessary condition for the achievement of economic success. This is difficult to measure because it can be observed only after a substantial time lag and because many departments other than R&D need to collaborate optimally to make this success happen. Thus, while it is difficult to measure the overall economic R&D success of an organization, it appears to be even more difficult to determine the appropriate share by which individual laboratories have contributed to success. Some of the literature puts trust in the ability of experienced managers to solve this problem by asking them to reveal their perceived evaluation of success for individual laboratories. However, the particular division of labor between laboratories and the dispersed customers for whom they work might involve specific biases, depending on whom one prefers to interview. Figure 18 classifies different approaches that one could take. Most frequently, R&D managers from headquarters (Box (4)) have been asked to evaluate the R&D performance of individual laboratories or the success of the overall R&D organization of their company. It is obvious that their answers could be biased in support of their own unit and could downgrade other units that they may view as competitors for scarce internal resources. A few other studies ask managers of foreign R&D units to evaluate its success from their point of view (Box (5)). This could stimulate an opposite bias. No study is known that would contrast evaluations from different viewpoints to check the validity of the responses. This leaves open a wide field for further studies. It is obvious that by comparing responses from interviewees who might be located at different boxes and by finding significant differences, one would immediately become aware of biases and problem areas that need further attention. Similar biases become obvious by comparing the views of R&D managers and scientists in the same laboratories (Compare Boxes (4) and (7) or (5) and (8)). Let us illustrate this point with reference to a strictly national study: in a study of German R&D laboratories, it was found that while managers thought that reducing budget overruns was the second most important issue in securing higher laboratory success, bench engineers in the same laboratories perceived that reducing delays in decision making was the second most important issue in this respect. This can have severe consequences: if such differences in the perception of issues cannot be resolved, many of the management efforts might well be futile. However, the limitations of research perspectives on international R&D cannot be resolved in short time, and, therefore, we issue an early warning to keep this limitation in mind.

Figure 18: Different perspectives on criteria to evaluate foreign R&D units

Viewpoint	Headquarters	Foreign unit under study	Other foreign units
General management	(1)	(2)	(3)
R&D management	(4)	(5)	(6)
Individual R&D scientist	(7)	(8)	(9)
R&D customers	(10)	(11)	(12)

7.2 First results

In earlier chapters we indicated that there are benefits and costs in internationalizing R&D activities. This leads to the question whether indicators of the outcome of an internationalization process could be identified or whether correlates of success of any given organizational structure could be found. As indicated in Chapter 7.1, it is difficult to collect such information because most of it is perceptional, which leads us to suspect biased viewpoints of those who answer questionnaires or participate in interviews. Some authors even go a step further and deny the possibility of success measurement altogether, as evidenced in this quotation: "Effectiveness and efficiency of R&D are not amenable as criteria for the evaluation of alternative configurations of international R&D networks. This is because of interdependencies with other influencing factors and substantial measurement problems" (Specht/Beckmann, 1997, p. 426). Such difficulties may explain why there is hardly any information available on R&D success. If success is reported at all, most of the information is collected from the headquarters of the respective companies. Only if that is not so, will we mention it in the following discussion.

Perceptions are mostly measured on five or seven point Likert-type scales. It will be mentioned explicitly if other types of measures are employed.

Before relating specific organizational structures of international R&D to success, we want to report on success measurement at an undifferentiated scale. The question here is whether a higher share of international R&D activities in a company is:

(1) related positively to effectiveness because the activities bring R&D closer to the markets; and
(2) related negatively to efficiency because of the higher efforts to coordinate and the relatively low importance put by management on savings in factor cost when choosing an internationalization strategy in R&D.

Minimizing total cost would indicate an optimal level of internationalization. The general idea might be depicted as in Figure 19. However, this might appear to be too high a level of abstraction.

Figure 19: Optimum level of internationalization

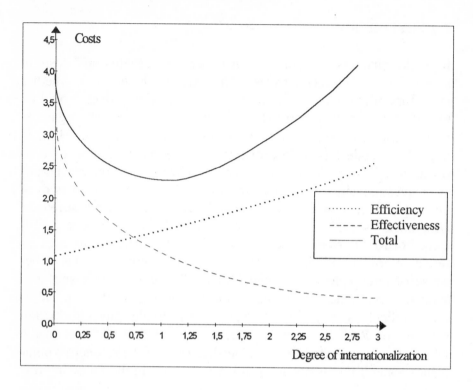

In a questionnaire study that was answered by the top management of 269 of the 780 largest German industrial companies, some embarrassing findings were made (Brockhoff, 1990) particularly with respect to the assumption of internationalization and effectiveness.

According to this study, a high percentage of R&D personnel employed abroad is associated with problems of ineffectiveness ($p<0.1$). This seems to indicate that either ineffective companies choose technology augmenting strategies by setting up international R&D units, which at the time of the interviews had not borne fruit yet, or it could mean that strategic planning in highly internationalized firms requires a coordination cost higher than its contribution to the results. This coordination effort may be even higher in cases where foreign subsidiaries have been acquired or where their management strives for more local autonomy. The size of the foreign unit, as measured by its R&D personnel, could lend support to this desire.

As hypothesized in Figure 19, efficiency is negatively correlated with the shares of R&D spent abroad. These shares correlate with overruns in development time ($p<0.09$) and with complaints about "bootlegging" ($p<0.01$). This indicates problems in controlling the activities.

These relationships are weak, and they explain only a small share of the variance. But they should be understood as warning signals. Conducting R&D abroad requires special attention to ensure high effectiveness and, to some degree, also high efficiency. This was mentioned above in the discussion of communication needs.

Effectiveness and efficiency may be operationalized by evaluating indicator variables. Thus, the question "To what extent did the results achieved by foreign R&D facilities during the past three years meet the expectations?" is assumed to be indicative of effectiveness. Responses could vary from 0% to 100%, and even beyond this mark. Overall, von Boehmer (1995) reports 75% as the median response to this question; responses from German respondents judging foreign R&D laboratories reach about 82% (Schmaul, 1995). We have found that expectations are met best in German (86.5%) and Japanese laboratories (81.8%), and least in French (70.3%), British (71.4%) and U.S. laboratories (72.1%); however, the differences are not significant.

Efficiency and effectiveness alike might be measured by the responses to the question: "What share of the total spending of the foreign R&D facility would you expect to find being wasted?" The same question was used in a few other studies on R&D efficiency to develop a notion of the reliability of the responses. The median response to this question is 11.6% for local German laboratories (Brockhoff, 1990), 10% for foreign laboratories of U.S., British and German firms (von Boehmer, 1995), and 15.2% for foreign laboratories of German firms (Schmaul, 1995). These differences are within a narrow range, and they are not significant. It appears that this question leads to reliable responses. Whether these responses are valid remains to be established.

Another overall measure of success is the share of money being wasted. This reflects perceptions on inefficiency and ineffectiveness. The highest shares of money perceived as being wasted are reported for laboratories in Great Britain (15.4%), the U.S. (15.3%) and France (14.7%), while the lowest values are achieved in Japan

(8.2%) and Germany (9.3%) (without considering countries that have not specifically been named); these values are significantly different from each other. They support the earlier findings on meeting expectations.

The differences between these measures should not be interpreted only as reflecting national cultures. The way that laboratories were set up as well as task assignments for the laboratories may vary by country; the degree of autonomy enjoyed by the laboratories may be different from one country to the other as well as the management influence by the headquarters. Too few cases are available in the studies to differentiate all these criteria and retain enough cases for statistical analyses.

It is also found that laboratories that were acquired for technological reasons meet expectations significantly better than those acquired for other reasons or those started from scratch (von Boehmer, 1995, p. 112). With respect to other overall success measures, the rankings of the share of money wasted in these laboratories reproduce exactly the same national rankings as mentioned before, although the differences are not significant.

A similar result is reported if laboratories are grouped by task assignments. Expectations are met best by "cooperating basic research laboratories" (82.5%) and "all-round R&D units with a local focus" (80.4%); the worst results are reported for the "all-round R&D units with a global focus" (69.6%) and the "local problem solvers" (69.7%); the "applied researchers" take a middle position (74.6%). One might speculate that the results depend on the clearness of the objectives set for the different types of laboratories as well as the degree of their centrality. The headquarters might distrust decentral laboratories, particularly if these are autonomous at the same time. Then, coordination is a difficult task.

As seen from the point of view of the foreign R&D establishments, R&D success does not seem to be substantially different between Asian, European, and U.S.-American locations (Kuwahara/Iwata, 1997). However, the responses are based on an "overall assessment of (the Japanese) company's R&D performance", and they use only three levels of success. Thus, there is not enough differentiation to evaluate success of foreign R&D establishments from their own perspective or for a cross-validated view.

Building on von Boehmer's experiences, a more elaborate scheme of success measurement was developed and applied to identify the success of different organizational arrangements for international R&D (Schmaul, 1995; Brockhoff/Schmaul 1996). In Table 6, the success criteria and the factors on which they load with more than a loading of .5 are shown. These include most of the criteria suggested in other research (Beckmann, 1997, p. 234).

Originally, 11 different criteria were chosen to describe success. It turned out that the respondents were unable to discriminate clearly between efficiency and effectiveness, although the standard descriptions of the terms (to do things right and to do the right things) had been added. Therefore, both criteria were aggregated into one construct. Thus, ten success criteria remain. Based on 67 observations, a principal component factor analysis was performed. It led to three distinguishable factors: meeting timing objectives, performance, and budgeting goals. These success criteria are frequently mentioned as being important in project management.

Based on the factor values, it was possible to discriminate between three clusters of laboratories: those 31% that exhibited above average success with respect to all factors, a group of 23% which reported less than average performance with respect to all three factors, and a large group (46%) of remaining laboratories with intermediate success. In this last group, the majority of the laboratories are perceived as underachieving with respect to timing and budgeting but as being successful in meeting the performance goals. Thus, they fail on two important criteria that co-determine market success while they appear to meet technical success criteria. However, the factor structure indicates that "success" requires the achievement of all three factors. There is little room for compensation between these.

Unfortunately, in von Boehmer's study success cannot be related to organizational structures. This has been achieved in further research. A summary of this is presented in Table 7.

Table 7 is partitioned into three major blocks of rows. The first block shows the success of the three types of laboratory structures. The results are significantly different with respect to timing and performance while budgeting is at the limit of acceptable significance. The hub model appears to be the most successful structure for internationally dispersed R&D laboratories. Timing and performance

Table 6: Success criteria and success factors

Questionnaire items	Factor 1: Timing	Factor 2: Performance	Factor 3: Budgeting	Commu-nality
What share of your total R&D budget would you expect to be wasted due to inefficiency or ineffective-ness?	-.857			.762
Share of projects started late?	-.766			.795
Share of projects meeting their time schedule?	+.702			.798
What is the average time overrun on projects?	-.496			.611
Average share of projects terminated before com-pletion?		-.923		.908
Average share of projects completed successfully?	+.543	+.710		.800
Average extent of budget overruns?			-.908	.847
Average share of projects that met almost exactly the planned budget?			+.710	.721
Share of projects that met almost completely the planned project goals?		+.536	+.589	.651
To what extent do the re-sults of the foreign R&D facilities meet your ex-pectations of efficiency and effectiveness?			+.537	.666
Variance explained Cronbachs alpha	52.4% .841	11.2% .769	12.0% .706	

Only factor loadings larger than .49 are shown.

Source: Schmaul, 1995.

goals are met more frequently than expected, and budgeting goals are met as in the other two structures. However, it requires substantial coordination efforts which are not reflected here, particularly if the number of spokes grows. Then, cross-information and cross-fertilization among those laboratories may be major problems. The hub model appears to make it easier for companies to explore more fundamental technological questions or cross-sectional technologies. Often, these are only of limited interest to the decision makers in subsidiaries or business units, and it is, therefore, the headquarters which develop support for such activities (Carbonare/Völker, 1996, p. 60). This could then be followed up on by central research laboratories which at the same time represent the hub.

Quite as expected from the high cost of its coordination, the network model performs weakly on timing and budgeting and achieves only average success with respect to technical performance. Disfunctional competition and unclear competence distributions could be the substantial weaknesses of this model. This might become more clear if the dispersed units were audited.

Although it is believed that there is a trend towards the network model (Gassmann/von Zedtwitz, 1998), its effective management appears to offer substantial challenges. The difficulties of establishing management systems that are compatible among all R&D sites, that allow for an intensive information flow between all units, that coordinate goal systems, and that provide for effective, unanimously accepted mechanisms for conflict resolution are particularly recognized in companies employing network models (Gassmann/von Zedtwitz, 1998). Such systems come at a high price. Therefore, the cost of coordination among the units might outweigh the potential benefits from tapping various sources of knowledge. Also, contrary to the hub model, the performance of more fundamental work might be at risk.

Very interestingly, a weakness in meeting performance goals of the competence centers is attributed to the competence model. The low performance evaluation is not compensated by any of the other two success criteria. It might be speculated that this reflects a response bias from the managers responsible for the headquarters' laboratories. They may feel deprived of their control of power. Alternatively, the headquarters may be too far away from the centers of competence to

Table 7: Success and organizational structures

Structure	Timing	Performance	Budgeting	Overall
Hub model	+	+	∅	+
Competence model	∅	-	∅	-
Network model	-	∅	-	∅
Significance	0.02	0.001	0.10	0.001
Autonomous facilities	-	n.r.	-	∅
Shared R&D decision making	∅	∅	∅	+
Headquarters dependent	n.r.	+	n.r.	n.r.
Locally dependent facilities	+	+	∅	+
Significance	n.s.	n.s.	0.001	n.s.
Degree of centralization	0.017	0.015	0.101	-
Significance	n.s.	n.s.	n.s.	-
Degree of autonomy	-0.135	-0.183	-0.190	-
Significance	n.s.	0.046	0.038	-

+, -, and ∅ indicate that relatively more laboratories than could be expected in the particular group have above average success, below average success, or average success with respect to the criteria given.
Figures indicate Kendall's tau correlation.
n.s. = not significant.
n.r. = no record, i.e. no difference between expected and observed values at any level of success.

Sources: Schmaul, 1995; Brockhoff/Schmaul, 1996.

adequately judge their success or failure. Companies that apply a "split and grow" policy, which supports new laboratories if members of older laboratories come up with specific and interesting product development ideas, may have let this go too far. Furthermore, the assignment of tasks to particular laboratories that made them centers of competence may well have been a "political" or negotiating process that failed to result in a successful structure. A final explanation relates the competence center to the not-invented-here-syndrome. If a certain laboratory has truly acquired specific competencies and receives strong feedback in this respect for a number of years, it might develop the not-invented-here-syndrome. Smaller laboratory teams do so over time if they feel they have developed particular knowledge

(Katz/Allen, 1982; Albach, 1993). Should this be true, competence centers would have to be rejuvenated from time to time. This rejuvenation could be achieved by redesigning the internal division of labor in performing technological development. This redesign might be called for anyhow, if the perceived level of competencies does not correspond to the evaluations the laboratory receives from its partners or customers within the firm. Also, systematically substituting a small share of the personnel at competence centers, either by exchange with other units or by introducing new staff from the outside, might support the process of rejuvenating. Finally, competence centers should enrich their own competencies by employing listening posts at major sites of knowledge generation in their or related fields; if the company employs other R&D units in those locations, it might be a good idea to "implant" these smaller listening units into these R&D organizations as this might further strengthen the ties between the different parts of the R&D organization. In yet another view, it should be kept in mind that competence centers develop technologies that will typically serve more than one business or business unit. The new knowledge might be received and understood differently by these internal customers. As this becomes known, suspicion could grow that customers are served differently by the competence center, which again initiates conflicts and rivalries.

A further differentiation needs to be made. If competencies are assigned to particular laboratories because of the size of a local market or a strong hierarchical position of their promotors (Gerybadze, 1997, p. 28), this might lead to substantially different success perceptions than if competencies are achieved in a more or less time-consuming learning process or have to be defended over and over again against competing laboratories.

Further consequences arise from the observations on performance measurement. It becomes even more clear that subjective measures need to be cross-validated by asking all parties involved; alternatively, objective measures would need to be developed which take the lags between the adoption of an organizational structure and the possible consequences into consideration.

Turning now to autonomy and centrality as two criteria to describe foreign R&D laboratories, it becomes obvious from the lowest part of Table 7 that a higher degree of autonomy is considered to be rather unsuccessful with respect to technical performance and to budgeting. No significance can be established with respect to timing. As the network structures are often found in the group of the autonomous laboratories, this result does not come as a surprise. The result is congruent with the observation of focusing, streamlining, bundling, or choosing narrowed-down strategic orientations, both at the laboratory level and at the level of international projects (Gerybadze, 1997, pp. 19, 25). These activities aim at achieving better co-ordination by reducing autonomy among dispersed R&D units without sacrificing the underlying network (or competence center) structures. Again, a respondent bias might explain the result. It has been argued that foreign R&D is "liable to be affected by high autonomy-control tension as a result of tension between different institutional environments at the functional level (research/corporate) and at the geographical level (host country/home country)" (Asakawa, 1996, p. 24). This type of tension is defined as a perceived discrepancy in the degrees of autonomy and control by headquarters and by local laboratories. This discrepancy could relate to the reality as well as to a normative, most wanted situation. Local R&D managers seek a certain degree of autonomy for their laboratories, and they apply interactions with the local scientific environment as one of the power bases to achieve this. In a similar sense, it is argued that: "Experiences with mounting coordination problems of globally dispersed and very often cooperative R&D projects have in the meantime led to new attempts at consolidation . . ." (Gerybadze, 1997, p. 19, similarly p. 25).

However, the obvious explanation of a possible bias is not too convincing for at least two reasons.

- Firstly, the centrality index is not related significantly to any of the success criteria. A bias could have meant that higher degrees of centrality would be considered significantly more successful by headquarters respondents.
- Secondly, turning to the middle part of Table 7, the locally dependent facilities receive far better evaluations than the headquarters' dependent facilities, although both are predominantly of the hub model type.

As expected from the aforementioned results, the autonomous facilities are again rated as performing rather unsuccessfully. This might explain a re-centralization trend that appears to be prominent among 24 multinational companies (Gassmann, 1997b, p. 61). With respect to all the success criteria, shared R&D decision-making facilities exhibit average results with a more than expected frequency, and they even achieve above average overall success evaluations relatively more frequently.

7.3 Differentiated success evaluation

Dissatisfaction with laboratory performance might arise from "a lack of a clear charter for each of the laboratories in the learning process" (De Meyer, 1993). This learning process is of considerable importance in explaining the existence of internationally dispersed R&D units. Adding this to the above observations might lead to a dynamic view of laboratory success. This is sketched in Figure 20. It is hypothesized that this process evolves over time. Whether it produces growth swings of a constant or declining amplitude will depend on the possibility of learning and managing the process.

The results reported above need to be interpreted more as signals than as facts. This has many reasons. The small sample size and some missing values contributed to problems of validating some of the expected relationships. For the same reasons, some cross relationships, like market orientation and the many types of laboratory models, could not be analyzed by statistical means. The problem of response bias cannot be resolved fully, as this would have necessitated interviewing the heads of foreign laboratories from the same companies as well. Other criteria, such as the personality traits of laboratory directors, type of task assignment, environmental conditions, or type of technological progress could not be controlled. For instance, overall results indicate that the share of funds considered to be wasted and the perceived level of inefficiency are positively correlated with the share of research expenditures and negatively correlated with the share of development expenditures. Geographical closeness to marketing or production facilities seems to favor success. Such observations could mean an interaction with organizational variables as well as a distrust in those laboratories that digress from day-to-day hands-on business.

A true and independent measure of market success could not be applied, as pretests showed that few laboratory managers could reliably estimate the average market success of their contributions. Interestingly, a Delphi-type approach may be helpful in eliciting reliable responses of this type (Bardenhewer, 1998). The dynamics of laboratory development cannot be reflected in a snapshot type evaluation of success. Again, longitudinal research is badly needed to correct this.

Figure 20: Dynamics of success and autonomy of laboratories

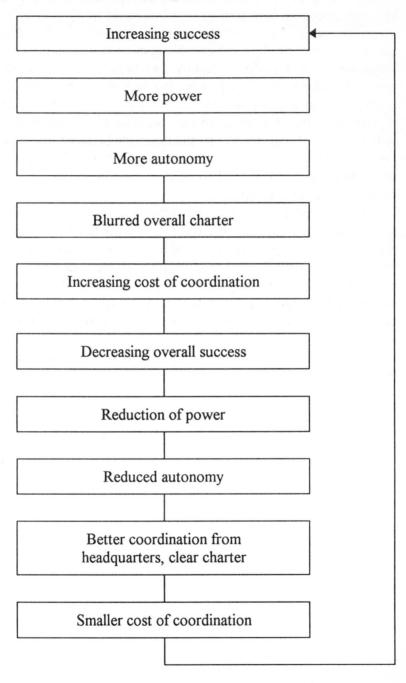

One might speculate that in setting up laboratories, the headquarters will chose, at first, a centralized decision making approach and give little autonomy to the foreign unit; upon observing success according to some or all of the criteria mentioned here, the foreign unit might be granted more autonomy and become less dependent on centralized decisions. It may even become less dependent on decisions passed locally but outside the laboratory. Then, the big question is: how does one determine an optimal level of autonomy in operational terms and with respect to the time lags between input and output of laboratory work? A reverse development may be necessary for acquired units, as these tend to be rather autonomous and decentral at the moment of acquisition.

The hub model appears to be rather successful. However, what will happen if the company decides to add more laboratory sites? Will the hub become largely unmanageable with an increasing number of "spokes", and will the choice of a different type, like the center of competence model, be only a partial relief as it could lead to a loss of overall technological performance? Evaluating the different types of costs and benefits involved in these choices is a difficult and yet unresolved task.

The present state of success evaluation is rather limited. Much more research is needed to guide managers in their decisions.

8. Summary and conclusions

A new phenomenon can be observed even in large, industrialized countries during recent years. A growing number of firms started spending mounting sums of money for R&D in foreign countries but not at their headquarters. This trend has alarmed politicians, and it has caused technology managers to review their strategies, particularly, if these were primarily based on the centralized generation of new knowledge. The trend reflects an important facet of growing internationalization. This trend is also represented by the employment of people of many nationalities in real or virtual project teams or in the participation of firms in international R&D cooperations. Disregarding such aspects, which cause management problems of their own, we concentrate on a situation where a firm operates not only one or many laboratories in one country, but also runs laboratories in nations which are considered foreign from the perspective of its headquarters. Then, a first question arises: What explains this phenomenon of internationalization of R&D laboratories?

During recent years, merger and acquisition activities have become more and more frequent. On the verge of these activities, laboratories were acquired. Only a few acquisitions aimed at getting access to a specific knowledge base. Consequently, many foreign laboratories were acquired incidentally. A major problem is then to formulate a technology strategy that integrates the newly acquired laboratories into the knowledge generating capacities of the specific firm. Strategy formulation is a consequence of the acquisition. If such a strategy fails to be developed, the incidentally acquired laboratories either have to be spun off or shut down. It goes without saying that closing a laboratory might be an expensive process, at least in those countries where employees are well protected by legal arrangements, codetermination laws, or a protective culture.

Other laboratories were established intentionally, either by mergers and acquisitions or by allocating resources to internal activities for this purpose. Thus, the existence of these laboratories ideally follows a technology strategy.

At least four different groups of reasons could be named to explain the strategic intent:

(1) Empirical research has established the existence of an optimal laboratory size. This is an employment level which maximizes perceived laboratory performance. If such a level exists, establishing new laboratories becomes relevant as soon as the optimum level of existing laboratories is surpassed.

The level varies with respect to the particular conditions by which technological knowledge is produced in different fields as well as with respect to cultural communication habits. Measurement of performance is crucial for the determination of the optimum, and at present this cannot be achieved very reliably. Increasing experiences with the performance measurement of laboratories should help to clarify the situation. The use of objective performance measures would be most welcome. This necessitates an analysis of outputs (such as patents) and outcomes (such as generation of additional sales or profits), and the contribution of R&D to their generation.

(2) Irrespective of the employment level that might constitute an optimum laboratory size, foreign laboratories might be established to improve R&D efficiency. Lower cost resources, higher productivity of R&D work, and shorter distances between sources and sinks of communication can all be cited as relevant to increasing efficiency. Efficiency can be measured at the laboratory level by relating R&D outputs to R&D inputs. This might explain why it has been paid substantial attention, particularly among controllers. It is difficult, though, to compare efficiency among laboratories that have different missions because the variables that define appropriate benchmarks need to vary with the missions.

(3) Efficiency improvements might be of less relevance than effectiveness improvements. Independent of the aforementioned reasons, foreign laboratories might contribute to enhanced effectiveness. On the one hand, this could be achieved by improving access to inputs, in particular new knowledge and, on the other hand, by sharpening the information on required outputs. In both cases, the neighborhood of the R&D laboratory to the informa-

tion sources facilitates personal communication, observation of activities, and interaction. These activities might be necessary to transfer information that cannot be transferred by communication media. The greater richness of the transfer is a necessary condition for the identification of relevant information and its use.

(4) Laboratories interact with their environment. Culture as an environmental factor might inhibit or facilitate the performance of certain types of R&D work. The laboratory's natural environment might be affected by exhausts or the non-aesthetic looks of its installations. Environmental conditions, like the availability of seawater, natural materials or favorable weather conditions, might be a benefit for certain locations for particular tasks. The artificial environment, as represented in legislation or economic conditions, might be another factor to be considered in choosing locations. Clearly, these conditions vary considerably around the globe.

A technology strategy has to weigh different reasons. This causes a number of problems because some reasons are not completely independent of each other. For instance, establishing a laboratory at a low cost location in a foreign country in order to maximize efficiency might at the same time contradict any attempts at keeping up high effectiveness because communication links between the laboratory and the other knowledge sources or the laboratory and its customers might become longer, more formal, and, thus, the communication itself less rich. Some other reasons, particularly some of those mentioned in (4), above, might establish binding constraints on laboratory locations. Thus, a constraint optimization which considers all reasons may be called for. This problem cannot be operationally formulated and solved because of fuzzy variables, unclear relations between inputs and goals, noticeable time lags, and substantial uncertainty. This might explain the co-existence of quite a number of different solutions used in the internationalization of R&D laboratories, even within one industry.

Furthermore, the relative importance of different reasons might shift over time. Then, the question arises whether laboratory missions need to be changed or whether more severe consequences, including closing down a laboratory, need to be drawn. Mission changes involve scientists and engineers, people who are trained and competent in special fields and who cannot typically achieve a comparable level of output if required to work in other fields. Mission changes will, therefore, most often involve personnel turnover or the integration of new personnel of a different type into an existing organization. Both are time-consuming and possibly also expensive tasks. Mission changes should, therefore, be expected to occur at a slow rate. Such changes can be observed in reality. However, no standard pattern of such changes has evolved.

Once laboratories have been established or acquired, the next question that arises is: who should make their strategic and operational decisions? This involves a complex set of variables, three of which have been found to be of particular relevance:

(1) Centrality of decision making, which measures whether decisions are made close to headquarters or further away from it, perhaps at a local unit. In case of low centrality, it is usually assumed that the locus of decision making is close to the laboratory in question. However, it is conceivable that centrality is low and decisions are not made at the laboratory or even at the location of the laboratory but at another subsidiary. Thus, centrality may refer to aspects that can be separated from the distance between the laboratory and the decision making unit. In turn, distance and communication frequency are interdependent and can, thus, influence the performance of laboratories.

(2) Autonomy is a construct that refers to various strategic and operational decisions that determine the tasks and the task performance of a laboratory and the degree to which these decisions are made at the laboratory or at another departmental unit of the firm. Fully autonomous laboratories make all decisions by themselves. Non-autonomous laboratories can be close to a decision making unit or far away from it. While autonomy has managerial implications for the choice of tasks and their performance (effectiveness and efficiency) or the choice of personnel, distance is a communication cost factor.

(3) A laboratory's degree of power can be independent from the aforementioned variables. This has given rise to the network power paradox. It means that autonomous laboratories might have a weak power base and that powerful laboratories might be autonomous as well. Power is determined by four variables: namely,

- criticality of a laboratory's work for the whole organization;
- substitutability of its work by other laboratories;
- interaction with other departments; and
- immediacy of the potential influences on the overall organization.

Construction of an overall power index on the basis of these variables has not yet been tried and tested. It is speculated that prominent structural models of laboratory organization such as the hub model, the center of competence model, or the network model might exhibit particular power distributions among themselves. In a descriptive sense, we frequently observe medium powered laboratories with relatively high levels of autonomy. In a normative sense, one might ask whether this observation coincides with the greatest success potentials and to what degree the high power of individual laboratories needs to be checked and balanced to avoid them become inefficient by attracting more resources than necessary.

A strict hierarchy of power among laboratories might facilitate the coordination of their activities. However, during recent years network models of laboratories or competence center models, where several laboratories are competence centers for different technologies without establishing a hierarchy, have become quite common. This establishes organizational interfaces.

Managing organizational interfaces is difficult. It involves choosing instruments that respond in a very special way to the particular characteristics of the interfaces. These characteristics are:
(1) the level at which the interface is observed,
(2) the reason for the existence of an interface,
(3) the type of interaction across the interface,
(4) the characteristics of the tasks to be performed.

According to these characteristics, instruments of interface management should be chosen. These could be full-time assignments to interface management or apply instruments which are complementary to the tasks to be performed. Particular choices of instruments can only be made on the basis of plausibility. Therefore, the performance of the instruments should be observed to provide data for eventually better choices and improvements. In that respect, great hopes are vested by some observers in modern communication media that facilitate communication over great distances at ever decreasing costs per message and mile. These characteristics could favor further international dispersal of laboratories. However, it is argued here that personal exchanges are of particular relevance in the laboratory environement for at least two reasons:

(1) A substantial part of the tasks generates and necessitates badly encodeable information, sticky information, or tacit knowledge. Its transfer is best achieved by personnel interaction.

(2) Different specializations of laboratories create asymmetric information. This could breed opportunistic behavior. To avoid detrimental effects of opportunistic behavior, contracts could be closed. This would lead to high transfer costs, and it is doubtful that in view of the particular work of laboratories, namely generating new knowledge, very efficient contracts could be formulated. Alternatively, trust could be built to balance and to avoid opportunistic behavior. This necessitates personal contacts and their frequent renewal. For both reasons, the use of media might be more or less restricted to data exchanges and operational decisions. Personal contacts are important for the more strategic issues, and they could grow even more important as more foreign laboratories are established or more interaction among laboratories becomes mandatory.

The management of foreign R&D laboratories has to consider a complex system of relationships that is characterized by an extremely large number of variables. Some of these are not very well operationalized. It would be of high interest, therefore, to develop a notion of the potential influences of the variables on performance or success of individual laboratories. As mentioned above, overall success can be most easily measured by asking knowledgeable individuals for a performance evaluation. This leads to two different sorts of problems. Firstly, the standpoint of the respondent might be biased. Secondly, success or performance are not standardized variables, such that a number of operationalizations are possible. These could be developed in three different directions:

(1) measurement of economical success,
(2) measurement of technical success,
(3) measurement of socio-psychological or personal success.

Most of the known performance evaluations of foreign R&D laboratories are given by top managers of the headquarters' laboratory. It is evident that this might bias responses, particularly if foreign laboratories enjoy high degrees of autonomy. It is with this caveat that results from empirical analyses are presented. These results seem to indicate that the network and the competence models apparently show weaker performance than the hub model, in particular that variant of the model where the foreign laboratory serves as the spoke to a headquarters' hub. The results indicate that the search for an optimum distribution of autonomy and power over dispersed foreign laboratories is extremely difficult. Actually, it appears that the distribution itself is in flux. For many years, more decentralization in combination with a more even distribution of autonomy and power was considered the most beneficial for a firm. More recently, a number of companies has started to recentralize some of the decisions and laboratories that were decentralized in previous years.

What is evident from the available analyses is that the management of a considerable number of foreign laboratories necessitates an extra level of coordination in the firm. Various models whereby coordination can be achieved at this level can be observed. Again, more studies are needed to come up with a reliable and valid result for the most appropriate solution.

The previous sentence leads to a final, methodological remark. This work shows very clearly that the problems of the management of foreign R&D laboratories are multi-facetted. This means that a large number of potential success factors could be considered. Empirical research, therefore, requires a large number of observations or a smaller number of observations which are in some way homogeneous. To date, it is difficult if not impossible to arrive at a sample that would meet these conditions. In particular, to balance respondent biases it is necessary to collect perceptional data from various strategically positioned respondents within an organization. Hopefully companies will cooperate with external researchers to collect such data that might lead to a better understanding of the successful management of international R&D laboratories within one firm.

Literature

Ajami, R., Arch, G., Cooperating to compete: using technology to link the multinational corporation and the country, International Journal of Technology Management, Vol. 5, 1990, pp. 165-177.

Albach, H., Culture and Technical Innovation. A Cross-Cultural Analysis and Policy Recommendations. Berlin (Walter de Gruyter) 1993.

Albach, H., Global competition among the Few, Swedish School of Economics and Business Administration, Research Report 40, Helsingfors 1997.

Allen, T. J., Managing the Flow of Technology, Cambridge, MA (MIT Press) 1977.

Arimura, S., Global R&D Management - The case study of Matsushita Electrical Co., Ltd. and Sony Corporation, CD-Rom, Papers Presented at PICMET '97, Portland, OR. 1997.

Asakawa, K., External-Internal Linkages and Overseas Autonomy-Control Tension: The Management Dilemma of the Japanese R&D in Europe, IEEE Transactions on Engineering Management, Vol. 43, 1996, pp. 24-32.

Bardenhewer, J., Zur Integration der industriellen Forschung in ihr Umfeld. Empirische Ergebnisse aus Europa und Japan und ein Versuch der Wirkungsmessung, Diss. Kiel 1997.

Beckmann, C., Internationalisierung von Forschung und Entwicklung in multinationalen Unternehmen. Aachen (Shaker) 1997.

Beckmann, C., Fischer, J., Einflußfaktoren auf die Internationalisierung von Forschung und Entwicklung in der deutschen Chemischen und Pharmazeutischen Industrie, Zeitschrift für betriebswirtschaftliche Forschung, Vol. 46, 1994, pp. 630-657.

Behrman, J. N., Fischer, W. A., Overseas Activities of Transnational Companies, Cambridge, MA. 1980.

Beise, M., Belitz, H., Internationalisierung von F&E multinationaler Unternehmen in Deutschland, in: Gassmann, O., von Zedtwitz, M., Internationales Innovationsmanagement, München (Vahlen Verlag) 1996, pp. 215-230.

Beise, M., Belitz, H., Trends in the Internationalization of R&D - the German Perspective, Manuscript, Mannheim (ZEW) 1998.

Berge, C., The Theory of Graphs and its Applications, London, New York (Methnen, John Wiley) 1966.

Bergen, S. A., R&D Management: Managing Projects and New Products, Cambridge, MA (Basil Blackwell) 1990.

Boutellier, R., Kloth, B., Bodmer, C. E., Neue Organisationsformen globaler Forschung und Entwicklung, Zeitschrift Führung + Organisation, Zfo, Vol.5, 1996, pp. 282-287.

Brockhoff, K., Decision Quality and Information, in: Witte, E., Zimmermann, H.-J., Empirical Research on Organizational Decision Making, Amsterdam (North Holland) 1986, pp. 249-265.

Brockhoff, K., Stärken und Schwächen industrieller Forschung und Entwicklung. Umfrageergebnisse aus der Bundesrepublik Deutschland. Stuttgart (Poeschel Verlag) 1990.

Brockhoff, K., Forschung und Entwicklung, Planung und Kontrolle, 4th ed., München/Wien (Oldenbourg Verlag) 1994.

Brockhoff, K., Schmaul, B., Organization, Autonomy, and Success of Internationally Dispersed R&D Facilities, IEEE Transactions on Engineering Management, Vol. 43, 1996, pp. 33-40.

Brockhoff, K., von Boehmer, A., Global R&D Activities of German Industrial Firms, Journal of Scientific & Industrial Research, Vol. 52, 1993, pp. 399-406.

Brockhoff, K., et al., Managing Interfaces, in: Gaynor, G.H., edt., Handbook of Technology Management, New York (McGraw Hill) 1996, chapter 27.

Buri, M., Suarez de Miguel, R., Walder, B., Forschung und Entwicklung in der Schweiz 1986, Bern (Bundesamt für Statistik) 1989.

Carbonare, B. D., Völker, R., Steuerung globaler F&E in der Pharmabranche, in: Gassmann, O., von Zedtwitz, M., Internationales Innovationsmanagement, München (Vahlen Verlag) 1996, pp. 57-72.

Cartwright, S., Cooper, C. L., The role of culture compatibility in successful organizational marriage, Academy of Management Executive, Vol. 7, 1993, 2/pp. 57-70.

Chakrabarti, A. K., Information Technology for R&D in Global Business, in: Dean, P. C., Karwan, K. R., edt., Global Information Systems and Technology: Focus on the Organization and its Functional Areas, Harrisburg, London (Idea Group Publishing) 1994, pp. 361-374.

Chakrabarti, A. K., Burton, J. S., Technological characteristics of mergers and acquisitions in the 1970's in the manufacturing industries in the U.S., Quarterly Review of Economics and Business, Vol. 23, 1983, 3/pp. 81-90.

Chakrabarti, A. K., Hauschildt, J., Süverkrüp, C., Does it Pay to Acquire Technological Firms? R&D Management, Vol. 24, 1994, pp. 47-56.

Chakrabarti, A. K., Souder, W. E., Technology, Innovation and Performance in Corporate Mergers: A Managerial Evaluation, Technovation, Vol. 6, 1987, pp. 103-114.

Chester, A. N., Aligning technology with business strategy, Research Technology Management, Vol. 37, 1994, 1, pp. 25-32.

Chiesa, V., Managing the Internationalization of R&D Activities, IEEE Transactions on Engineering Management, Vol. 43, 1996, pp. 7-23.

Cohen, W. M., Levinthal, D. A., Absorptive Capacity: A New Perspective on Learning and Innovation, Administrative Science Quarterly, Vol. 35, 1990, pp. 128-152.

Colberg, W., Internationale Präsenzstrategien von Industrieunternehmen, Kiel (Vauk Verlag) 1989.

Cordell, A. J., Innovation, the multinational corporation: Some implications for national science policy, Long Range Planning, 1973, Sept., pp. 22-29.

Dahlstrand, Å. L., Growth and inventiveness in technology-based spin-off firms, Research Policy, Vol. 26, 1997, pp. 331-344.

De Meyer, A., Nestlé S.A., INSEAD Case, Fontainebleau 1988.

De Meyer, A., Technology Strategy and International R&D Operations, INSEAD Working Paper 89/62, Fontainebleau 1989.

De Meyer, A., Internationalizing R&D Improves a Firm's Technical Learning, Research.Technology Management, Vol. 42, 1993, pp. 42-49.

De Meyer, A., Mizushima, A., Global R&D Management, R&D Management, Vol. 19, 1989, pp. 135-146.

Deutsche Bundesbank, Technologische Dienstleistungen in der Zahlungsbilanz, Sonderveröffentlichung, Frankfurt (Deutsche Bundesbank) May 1996.

Dunning, J. H., Narula, R., The R&D Activities of Foreign Firms in the United States, International Studies of Management & Organization, Vol. 25, Spring/Summer 1995, pp. 39-74.

Dusters, G., Hagedoorn, J., Internationalization of corporate technology through strategic partnering: an empirical investigation, Research Policy, Vol. 25, 1996, pp. 1-12.

Erickson, T. J., Worldwide R&D Management: Concepts and Applications, Columbia Journal of World Business, Vol. 25, 1990, pp. 8-13.

Ernst, H., Patentinformationen für die strategische Planung von Forschung und Entwicklung, Wiesbaden (Deutscher Universitätsverlag) 1996.

Farris, G. F., Ellis, L. W., Managing Major Change in R&D, Research Technology Management, 1990, Jan.-Feb./pp. 33-37.

Fischer, W. A., Scientific and Technical Information and the Performance of R&D Groups, TIMS Studies in Management Science, Vol. 15, Amsterdam 1980, pp. 67-89.

Florida, R., The Globalization of R&D: Results of a survey of foreign-affiliated R&D laboratories in the USA, Research Policy, Vol. 26, 1997, pp. 85-103.

Florida, R., Kenney, M., The Globalization of Japanese R&D: The economic geography of Japanese R&D investment in the United States, Economic Geography, Vol. 70, 1994, pp. 344-369.

French, J. R. P. Jr., Raven, B., The Bases of Social Power, in: Cartwright, D., Studies in Social Power, Ann Arbor, MI 1959, pp. 150-167.

Gassmann, O., F&E-Projektmanagement und Projekte länderübergreifender Produktentwicklung, in: Gerybadze, A., Meyer-Krahmer, F., Reger, G., Globales Management von Forschung und Innovation, Stuttgart (Schäffer-Poeschel) 1997(a), pp. 133-173.

Gassmann, O., Internationales F&E-Management, München (Oldenbourg) 1997(b).

Gassmann, O., von Zedtwitz, M., Ein Referenzrahmen für das internationale Innovationsmanagement, in: Gassmann, O., von Zedtwitz, M., Internationales Innovationsmanagement, München (Vahlen Verlag) 1996, pp. 3-15.

Gassmann, O., von Zedtwitz, M., Towards the Integrated R&D Network - New Aspects of Organizing International R&D, Working Paper, 1998.

George, V. P., Globalization through interfirm cooperation: technological anchors and temporal nature of alliances across geographical boundaries, International Journal of Technology Management, Vol. 10, 1995, pp. 131-145.

Gerpott, T. J., Globales F&E-Management, Die Unternehmung, Vol. 44, 1990, pp. 226-246.

Gerybadze, A., Globalisierung von Forschung und wesentliche Veränderungen im F&E-Management internationaler Konzerne, in: Gerybadze, A., Meyer-Krahmer, F., Reger, G., Globales Management von Forschung und Innovation, Stuttgart (Schäffer-Poeschel) 1997, pp. 17-37.

Gibson, D. V., Rogers, E. M., R&D Collaboration on Trial, Boston, MA (Harvard Business School Press) 1994.

Graham, M. B. V., Pruitt, B. H., R&D for Industry. A Century of Technical Innovation at Alcoa. Cambridge (Cambridge University Press) 1990.

Granstrand, O., Fernlund, I., Coordination of Multinational R&D: A Swedish Case Study, R&D Management, Vol. 9, 1978, pp. 1-7.

Granstrand, O., Håkanson, L., Sjölander, S., Internationalization of R&D - a survey of some recent research, Research Policy, Vol. 22, 1993, pp. 413-430.

Hagedoorn, J., Strategic technology partnering during the 1980s: trends, networks, and corporate patterns in non-core technologies, Research Policy, Vol. 24, 1995, pp. 207-231.

Håkanson, L., Locational Determination of Foreign R&D in Swedish Multinationals, in: Granstrand, O., Håkanson, L., Sjölander, S., edts., Technology Management and International Business, Chichester (Wiley) 1992, pp. 97-116.

Håkanson, L., Nobel, R., Foreign Research and Development in Swedish Multinationals, Research Policy, Vol. 22, 1993, pp. 373-396.

Håkanson, L., Zander, U., International Management of R&D: The Swedish Experience, R&D Management, Vol. 18, 1988, pp. 217-226.

Hansen, T., Companies' Adaptation to Host-Government Research and Development Requirements: The Foreign Oil Companies under the Norwegian Technology Agreements, PhD. Diss. Kiel 1997.

Harpaz, I., Meshoulam, I., Interorganizational Power in High Technology Organizations, The Journal of High Technology Management Research, Vol. 8, 1997, pp. 107-128.

Haspelagh, P. C., Jemison, D. B., Acquisitions - myths and reality, Sloan Management Review, Vol. 28, 1987, pp. 53-58.

Hewitt, G., Research and Development Performed Abroad by U.S. Manufacturing Multinationals, Kyklos, Vol. 33, 1980, pp. 308-327.

Hölsä, T., Internationalization of R&D in Finnish Multinational Companies, Paper presented at the Finnish-Swedish Seminar on Internationalization of R&D, Helsinki 1994.

Honko, J., Internationale Vergleiche der Stärken und Schwächen der Innovationstätigkeit einiger Industrieländer, Zeitschrift für Betriebswirtschaft, Vol. 80, 1990, pp. 1315-1339.

Hood, N., Young, S., US Multinational R&D, Multinational Business, 1982, 2/pp. 10-23.

Hoppe, M. H., The Effects of National Culture on the Theory and Practice of Managing R&D Professionals Abroad, R&D Management, Vol. 23, 1993, pp. 313-326.

Hough, E. A., Communication of Technical Information Between Overseas Markets and Head Office Laboratories, R&D Management, Vol. 3, 1972, pp. 1-5.

Hounshell, D. A., Smith, J. K., Jr., Science and Corporate Strategy. Research and Development at Du Pont 1908 to 1980. Cambridge (Cambridge University Press) 1989.

Howells, J., The location and organisation of research and development: New horizons, Research Policy, Vol. 19, 1990, pp. 133-146.

Jaffe, A. B., Characterizing the "technological position" of firms, with application to quantifying technological opportunity and research spillovers, Research Policy, Vol. 18, 1989, pp. 87-97.

Katz, R., Allen, T. J., Investigating the Not Invented Here (NIH) Syndrome: A Look at the Performance Tenure and Communication Patterns of 50 R&D Project Groups, R&D Management, Vol. 12, 1982, pp. 7-19.

Krogh, L. C., 3M's International Experience, in: Khalil, T. M., Bayraktar, B. A., edts., Management of Technology II, Proceedings of the Second International Conference on Management of Technology, Miami. Industrial Engineering and Management Press, Norcoss GA 1990, pp. xxxiii-xxxix.

Kuemmerle, W., Building Effective R&D Capabilities Abroad, Harvard Business Review, Vol. XX, March/April 1997, pp. 61-70.

Kuemmerle, W., Optimal Scale for Research and Development in Foreign Environments - An Investigation into Size and Performance of Research and Development Laboratories Abroad, Research Policy, to appear.

Kuwahara, S., Iwata, S., An Analysis of Overseas R&D Activities of Japanese Manufacturing Firms, Manuscript, PICMET 1997.

Levitt, T., The Globalization of Markets, Harvard Business Review, Vol. 61, 1983, pp. 92-102.

Linstone, H. A., Turoff, M., edts., The Delphi Method. Techniques and Applications. Reading, MA (Addison-Wesley Publ. Comp.) 1975.

Lundgreen, P., Engineering Education in Europe and the U.S.A., 1750-1930: The Rise to Dominance of School Culture and Engineering Professions, Annals of Science, Vol. 47, 1990, pp. 33-75.

Mansfield, E., Teece, D., Romeo, A., Overseas Research and Development by US-Based Firms, Economica, Vol. 46, 1979, pp. 187-196.

Maringer, A., Ist Forschung und Entwicklung in Japan billiger? Die Betriebswirtschaft, Vol. 50, 1990, pp. 789-800.

McNichols, R., Communication in an International Environment, in: Gaynor, G.H."G.", Handbook of Technology Management, New York (McGraw Hill) 1996, pp. 35.1-35.10.

Medcof, J. W., A taxonomy of internationally dispersed technology units and its application to management issues, R&D Management, Vol. 27, 1997(a), pp. 301-318.

Medcof, J. W., Strategic Contingencies and Power in Networks of Internationally Dispersed R&D Facilities, Man., Academy of Management Annual Meeting, Boston, MA 1997(b).

Ministery of Trade, and Industry of Japan, Foreign operations by Japanese Companies, 24th. ed., Tokyo (Publ. Dept. of the Ministery of Finance) 1995.

National Science Board, Science & Engineering Indicators 1996, Washington, DC (U.S. Government Printing Office) 1996.

Patel, P., Are Large Firms Internationalizing the Generation of Technology? Some New Evidence, IEEE Transactions of Engineering Management, Vol. 43, 1996, pp. 41-47.

Pausenberger, E., Technologiepolitik internationaler Unternehmen, Zeitschrift für betriebswirtschaftliche Forschung, Vol. 34, 1982, pp. 1025-1054.

Pearce, R. D., The Internationalization of Research and Development by Multinational Enterprises, New York (St. Martins Press) 1989.

Pearce, R. D., Singh, S., The Overseas Laboratory, in: Casson, M., Global Research Strategy and International Competitiveness, Oxford (Blackwell) 1991, pp. 183-212.

Pearce, R. D., Singh, S., Internationalization of Research and Development among the World's Leading Enterprises: Survey Analysis of Organisation and Motivation, in: Granstrand, O., Håkanson, L., Sjölander, S., edts., Technolgy Management and International Business, Chichester (Wiley) 1992, pp. 137-162.

Perlmutter, H. V., The Tortuous Evolution of the Multi-National Corporation, Columbia Journal of World Business, Vol. 4, 1969, pp. 9-18.

Perrino, A. C., Tipping, J. W., Global Management of Technology, Research-Technology Management, 1989, May-June/pp. 12-19.

Pieper, U., Vitt, J., Die Messung der technologischen Verwandtschaft von Akquisitionsunternehmen, Manuscript, Kiel 1998.

Porter, M., Edt., Globaler Wettbewerb und Strategien der neuen Internationalisierung, Wiesbaden (Gabler Verlag) 1989.

Porter, M., The Competitive Advantage of Nations, New York (The Free Press) 1990.

Reddy, P., Sigurdson, J., Strategic location of R&D and emerging patterns of globalization: the case of Astra Research Centre India, International Journal of Technology Management, Vol. 14, 1997, pp. 344-361.

Reger, O., Mechanismen zur Koordination von Forschung und Innovation im internationalen Unternehmen, in: Gerybadze, A., Meyer-Krahmer, F., Reger, G., Globales Management von Forschung und Innovation, Stuttgart (Schäffer-Poeschel) 1997, pp. 82-132.

Reichwald, R., et al., Telekooperation. Verteilte Arbeits- und Organisationsformen, Berlin et al. (Springer) 1998.

Ronstadt, R. C., Research and Development Abroad by U.S. Multinationals, New York, London 1977.

Rüdiger, M., Vanini, S., Das Tacit knowledge-Phänomen und seine Implikationen für das Innovationsmanagement, Die Betriebswirtschaft, Vol. 58, 1998, to appear.

Scherer, F. M., Industrial Market Structure and Economic Performance, Chicago, IL 1971.

Schirm, K., Die Glaubwürdigkeit von Produkt-Vorankündigungen, Wiesbaden (Deutscher Universitätsverlag) 1995.

Schmaul, B., Organisation und Erfolg internationaler Forschungs- und Entwicklungseinheiten, Wiesbaden (Deutscher Universitätsverlag) 1995.

Schulz-Hardt, S., Lüthgens, C., Internationale Unterschiede in Risikoeinstellungen. Erklären sie das deutsche Zögern bei den Zukunftstechnologien?, in: Pinkau, K., Stahlberg, C., Technologiepolitik in demokratischen Gesellschaften, Stuttgart (S. Hirzel) 1996, pp. 76-95.

Seiffert, U., Globaler Wettbewerb wird bei F+E schon bald ähnlich scharf wie im industriellen Bereich, Handelsblatt, Oct. 16, 1990, p. 24.

Sommer, R., Sony's Innovationsmanagement, Köln (Sony) 1990.

Steenkamp, J. B. E. M., Baumgartner, H., Assessing Measurement Invariance in Cross-National Consumer Research, Working Paper, Dep. of Toegepaste Economische Wetenschappen, Katholieke Universiteit Leuven 1996.

Steenkamp, J. B. E. M., Ter Hofstede, F., Wedel, M., A Cross-National Investigation into the Individual and Cultural Antecedents of Consumer Innovativeness, Working Paper 9720, Dep. of Toegepaste Economische Wetenschappen, Katholieke Universiteit Leuven 1997.

Stevens, C., Technoglobalism vs. Technonationalism: The Corporate Dilemma, Columbia Journal of World Business, Fall 1990, pp. 42-49.

Stock, G. N., Greis, N. P., Dibner, M. D., Parent-Subsidiary Communication in International Biotechnology R&D, IEEE Transactions on Engineering Management, Vol. 43, 1996, pp. 56-68.

Süverkrüp, C., Internationaler technologischer Wissenstransfer durch Unternehmensakquisitionen, Frankfurt et al. (Lang) 1992.

Taggart, J. H., Determinants of Foreign R&D Locational Decision in the Pharmaceutical Industry, R&D Management, Vol. 21, 1991, pp. 229-240.

Teichert, T. A., Erfolgspotential internationaler F&E-Kooperationen, Wiesbaden (Deutscher Universitäts Verlag) 1994.

van Rossum, W., Comparison of Virtual R&D Laboratories with Traditional Corporate R&D, Manuscript, University of Groningen, NL 1998.

von Boehmer, A., Information Systems for Global Technology Management, in: Dean, P.C., Karwan, K.R., Edt., Global Information Systems and Technology: Focus on the Organization and its Functional Areas, Harrisburg, London (Idea Group Publishing) 1994, pp. 345-360.

von Boehmer, A., Internationalisierung industrieller Forschung und Entwicklung. Typen, Bestimmungsgründe und Erfolgsbeurteilung, Wiesbaden (Deutscher Universitätsverlag) 1995.

Vorort (Schweizerischer Handels- und Industrie-Verein), Forschung und Entwicklung in der Schweizerischen Privatwirtschaft 1996, Zürich (Vorort) 1998.

Weatherall, A., Nunamaker, J., Introduction to Electronic Meetings, Chandlers Ford 1995.

Yoshihara, H., Iwata, S., Results of the Questionnaire on the "Overseas R&D of Japanese Companies", Manuscript, PICMET 1997.

Zaininger, K. H., Aspects of Global Management of R&D Resources, in: Khalil, T. M., Bayraktar, B. A., edts., Management of Technology II, Proceedings of the Second International Conference on Management of Technology, Miami. Norcoss GA (Industrial Engineering and Management Press) 1990, pp. 269-279.

Zander, I., Technological diversification in the multinational corporation - historical evolution and future prospects, Research Policy, Vol. 26, 1997, pp. 209-227.

List of Figures

144

List of Tables

New in Management Science

T. Reichmann
Controlling
Concepts of Management Control, Controllership, and Ratios

1997. XIV, 338 pages. 174 figures.
Hardcover DM 128,–
ISBN 3-540-62722-7

Presented is a concise concept for the design of a ratio and management report system for each functional part of the company. The book addresses practitioners who seek decision support in their day-to-day business as well as scientists and students who want to inform themselves about the state of the art of controlling.

Please order from
Springer-Verlag Berlin
Fax: + 49 / 30 / 8 27 87- 301
e-mail: orders@springer.de
or through your bookseller

Errors and omissions excepted.
Prices subject to change without notice.
In EU countries the local VAT is effective.

K. Brockhoff
Industrial Research for Future Competitiveness

1997. XVI, 150 pages. 20 figures. 13 tables. Hardcover DM 85,–
ISBN 3-540-62842-8

Industrial research has come under pressure. Will recent budget cuts reduce competitiveness? Based on interviews in Japanese and European high-tech firms it is shown that research supports important potentials. These can be used for project funding, location decisions, and an analysis of sufficient conditions for research success. Careful management of the potentials should improve future competitiveness, and it should help to understand why industrial firms benefit from research and how.

Y.-K. Kwok
Mathematical Models of Financial Derivatives

1998. Approx. 250 pages. 40 figures. Hardcover DM 128,–
ISBN 981-3083-25-5

Until now, there was a lack of texts stressing the mathematical aspects of derivative pricing and so meeting the demands of students enrolled in the new mathematical and computational finance degree programs. *Mathematical Models of Financial Derivatives* fills this gap: it models derivative products based mainly on the differential equation approach, together with numerical solution techniques when appropriate. Research results and concepts are made accessible to the student through extensive, well thought out exercises at the end of each chapter.

L. Bianco, P. Dell'Olmo, A.R. Odoni (Eds.)
Modelling and Simulation in Air Traffic Management

(Transportation Analysis.)
1997. XII, 202 pages. 44 figures. 26 tables. Hardcover DM 98,–
ISBN 3-540-63093-7

Dealing with a wide range of topics and covering different aspects of current importance in ATM, the papers place particular emphasis on automation and application of mathematical models and computational algorithms for ATM systems. The volume thus offers readers a summary of recent progress in such important areas as new operational concepts for automated ATM, evolution of traffic characteristics, ground-holding algorithms, ATC simulation facilities and a number of other aspects of ATC flow management.

Springer

Springer-Verlag, P. O. Box 14 02 01, D-14302 Berlin, Germany.

Jak.4418/MNT/E/1

Springer
and the
environment

At Springer we firmly believe that an
international science publisher has a
special obligation to the environment,
and our corporate policies consistently
reflect this conviction.
We also expect our business partners –
paper mills, printers, packaging
manufacturers, etc. – to commit
themselves to using materials and
production processes that do not harm
the environment. The paper in this
book is made from low- or no-chlorine
pulp and is acid free, in conformance
with international standards for paper
permanency.

 Springer